Images of the Cosmos

Barrie W. Jones

Robert J. A. Lambourne

David A. Rothery

Hodder & Stoughton

in a[...]rsity

This book was produced during the preparation of the Open University course *S281 Astronomy and planetary science*. The course was produced by a team (listed at the end of the book), many of whom made contributions to this book. We would particularly like to thank Cheryl Newport, John Greenwood, Dick Sharp, Lesley Passey and Sue Dobson for the considerable and essential roles that they played. Thanks are also due to Carol Forward for word-processing the text.

BWJ RJAL DAR

Cover images
Nebulae in the Rho Ophiuchi region: © Royal Observatory Edinburgh/Anglo Australian Telescope Board; photograph by David Malin.
Saturn, taken on 26 August 1990 by the Hubble Space Telescope: © NASA; image supplied by the Space Telescope Science Institute, Baltimore, USA.
Junction of Heaven and Earth, by Camile Flammarion; image supplied by the Mary Evans Picture Library.

British Library Cataloguing in Publication Data
A record has been registered with The British Library

ISBN 0 340 60065 9

First published 1994
Impression number 10 9 8 7 6 5 4 3 2 1
Year 1998 1997 1996 1995 1994

Edited, designed and typeset by The Open University.
Printed in Singapore by Craft Print (Europe) Ltd.
This text forms part of an Open University Second Level Course. If you would like a copy of *Studying with The Open University*, please write to the Central Enquiry Service, PO Box 200, The Open University, Walton Hall, Milton Keynes, MK7 6YZ. If you have not enrolled on the Course and would like to buy this or other Open University material, please write to Open University Educational Enterprises Ltd, 12 Cofferidge Close, Stony Stratford, Milton Keynes, MK11 1BY, United Kingdom.

1.1

Foreword

Professor Arnold Wolfendale FRS, Astronomer Royal

It is an interesting property of the human brain that, long after we have forgotten numerical values, and names, and mathematical equations, we remember images – 'once seen, never forgotten' is a common phrase, and it certainly applies to many of the most impressive images shown here. The very professional Open University team have produced a balanced and exciting image-tour through the Universe for our benefit.

Images of the Cosmos is an invaluable aid to students of the Open University course on astronomy and planetary science, but its appeal will be wider than that – many others will be fascinated by the material within, both illustrations and text. Professionals, too, will benefit from its perusal, not least those working in the classical areas of the subject who will be able to see views taken through the 'new' windows of radio, infrared, ultraviolet, X-rays and gamma rays; all are represented here.

Taken all together, then, this volume will fill a significant niche and I firmly predict that it will go on to an astronomically large number of editions.

List of Plates
Part 1
The stars and the interstellar medium

Part 2
Our planetary system

Part 3

Galaxies and the Universe

Introduction

This book takes you on a journey through the cosmos, to visit the planets, stars and galaxies that make up the Universe. During this journey you will see images of astonishing beauty and colour, obtained in recent years by the best telescopes on Earth, and by spacecraft launched into the void beyond our atmosphere.

You will explore the Sun – a fairly typical sort of star – and a host of other stellar types, including the weird black holes. You will see that the space between the stars is not empty, but everywhere contains gas and dust, with which the stars have an intimate relationship.

You will explore the companions of the Sun, the planets and smaller bodies that make up the Solar System – worlds of extraordinary variety.

You will travel through the Milky Way – our galaxy – made up of thousands of millions of stars, and on to the vast numbers of galaxies beyond our own.

This is a journey that takes you not only through space, but also back through time, to the origin of the Solar System and even earlier, to the origin of the Universe itself in the Big Bang.

The journey is a feast for the eyes, yet eyes alone could never reveal all that you will see in the following pages. Marvellous though they are, our eyes are extremely limited. The light that they detect is just one small part of the wide range of *electromagnetic radiation*. Other parts of this range, such as X-rays and radio waves, also provide stunning images of the cosmos.

Electromagnetic radiation takes the form of a wave that consists of a fluctuating pattern of electric and magnetic fields that travel through empty space at an astonishing 300 thousand kilometres per second. The fields vary along the wave, passing through peaks and troughs of strength. The pattern can be very smooth, and repetitive.

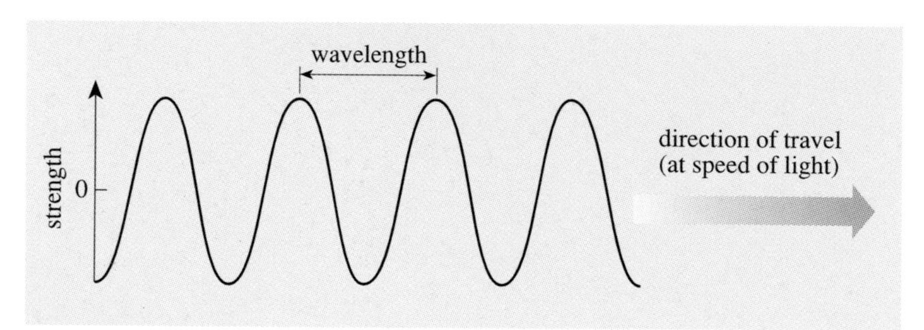

A smooth repetitive wave.

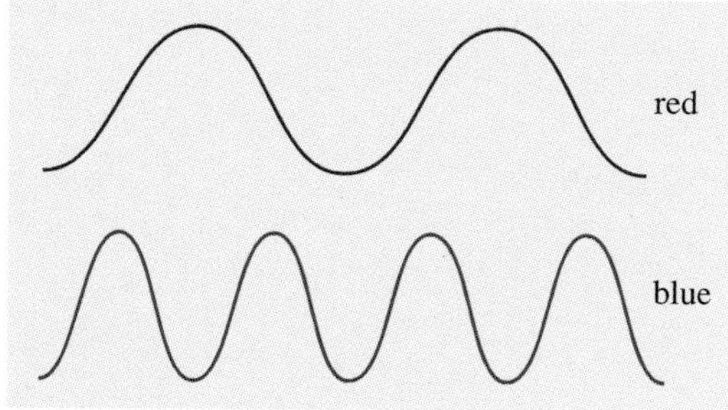

Light waves of different wavelengths, and the associated colours.

For a smooth, repetitive wave the distance from one peak to another is called the *wavelength*, and the associated rate at which the peaks pass a particular point in space is called the *frequency*. Each kind of electromagnetic wave – light, X-rays, radio waves and so on – corresponds to a particular range of wavelengths (or frequencies).

The visible range (light) runs from a wavelength of about 0.4 micrometres, to about 0.8 micrometres (a micrometre is a millionth of a metre). The shortest visible wavelengths produce the sensation we call 'violet'. Longer wavelengths produce blue, green, yellow, orange, until, with red, we come to the longest wavelengths that our eyes can detect.

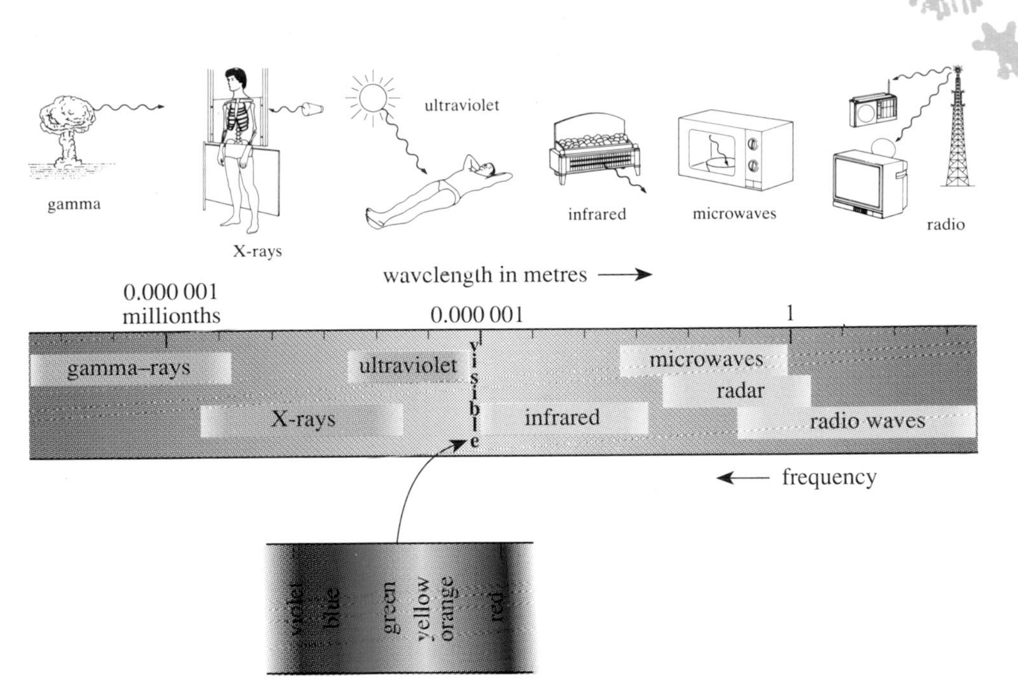

The whole spectrum of electromagnetic waves. The visible range (light) accounts for only a tiny part of the spectrum.

At yet longer wavelengths we come first to infrared waves, which our skin can detect as heat, then microwaves, including wavelengths used in microwave ovens, and then to radio waves, used not just by radios but also by television. Going the other way, to wavelengths shorter than violet, we come first to ultraviolet rays, among which are wavelengths that cause our skin to tan or 'burn'. Then we meet X-rays, widely used in medicine to peer inside our bodies, and finally, at the very shortest wavelengths, to gamma rays, which are emitted in nuclear reactions.

In the cosmos all of these various waves are emitted in a great range of important processes. So, in a very real sense, we see the cosmos hardly at all with our eyes.

The good news is that special detectors have been developed that enable us to 'see' beyond the visible. The bad news is that the Earth's atmosphere is opaque to gamma rays, X-rays, most ultraviolet rays, and to much of the infrared. Thus our detectors of these wavelengths have to be carried above the atmosphere if we are to see the invisible cosmos.

Let's turn now to the cosmos itself. It's BIG! The sizes and distances are beyond our everyday experience, and they also cover a very large *range* of values. Therefore in this book we have taken care, wherever possible, to attach a *scale* to each image, and to give the distance of the object from us. In some cases the scale and distance are measured by the comparatively familiar kilometre (nearly two thirds of a mile). In other cases they are given in less familiar terms – we sometimes use *light years* for distance, and *degrees* for size.

We use light years in order to reduce the numbers that we have to write down. The Earth's diameter is 12 756 kilometres, so the kilometre is a suitable unit here. It is also suitable for the Sun's diameter (1392 000 kilometres), but is only marginally suitable for the larger distance between the Earth and Sun (150 000 000 kilometres). The distance to the nearest star beyond the Sun is an unwieldly 39 900 000 000 000 kilometres, which we really do need to express in terms of a larger unit. We use the light year: the distance to the nearest star is only 4.22 of them! The light year is the distance that light travels through space in a year. Light streaks along, covering a distance of about seven times around the Earth every second, and there are a lot of seconds in a year, so one light year is a very long way. It is thus well suited for measuring large distances.

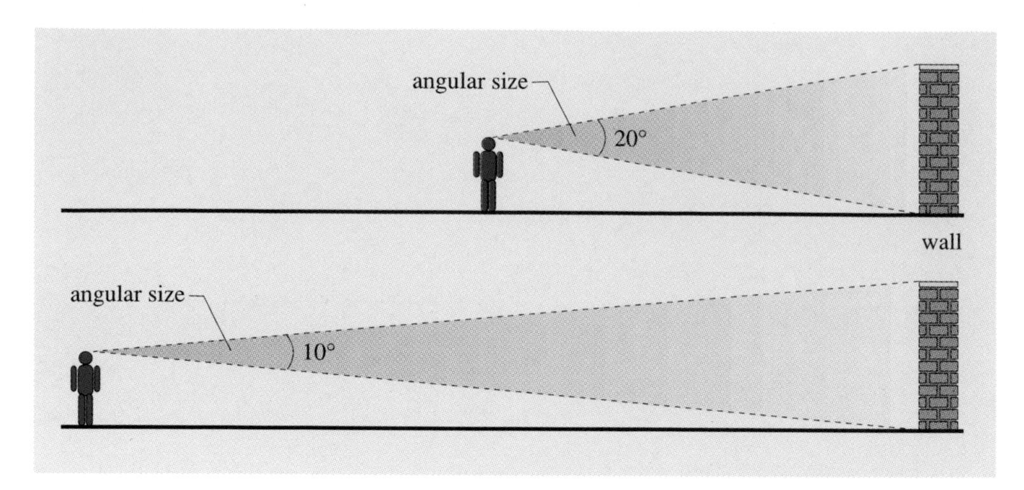

The angular size of the wall depends on the distance from which the observer views it.

We use degrees for size when we want to show how large an object appears to us from our Earthbound viewpoint. Degrees (°) measure angles, where 360° corresponds to one complete turn. You can see that the angular size of a wall, as measured from a point of observation, depends not only on the wall's height, but also on its distance from the point of observation: the greater the distance, the smaller the angle.

Celestial objects are far larger than walls, but they are at such *enormous* distances that the angles are usually very small. For example, the Moon and the Sun each has a comparatively large angular size among celestial bodies, yet each is only about half a degree across – far smaller than the angles in the diagram, and roughly equal to the angular size of a garden pea held at arm's length. When quoting small angular sizes, we often use subdivisions of the degree, namely the minute of arc, abbreviated to arcmin, and the second of arc, arcsec; there are 60 arcmin in a degree, and 60 arcsec in an arcmin. Thus, just as there are 60 seconds of time in a minute, and 60 minutes of time in an hour, so there are 60 seconds of angle in a minute of angle, and 60 minutes of angle in a degree.

It is a strange coincidence that the angular diameters of the Sun and the Moon are very nearly the same. The Sun's *linear diameter* is about 400 times that of the Moon, but it is also about 400 times farther away, so their *angular diameters* are about the same, as illustrated here.

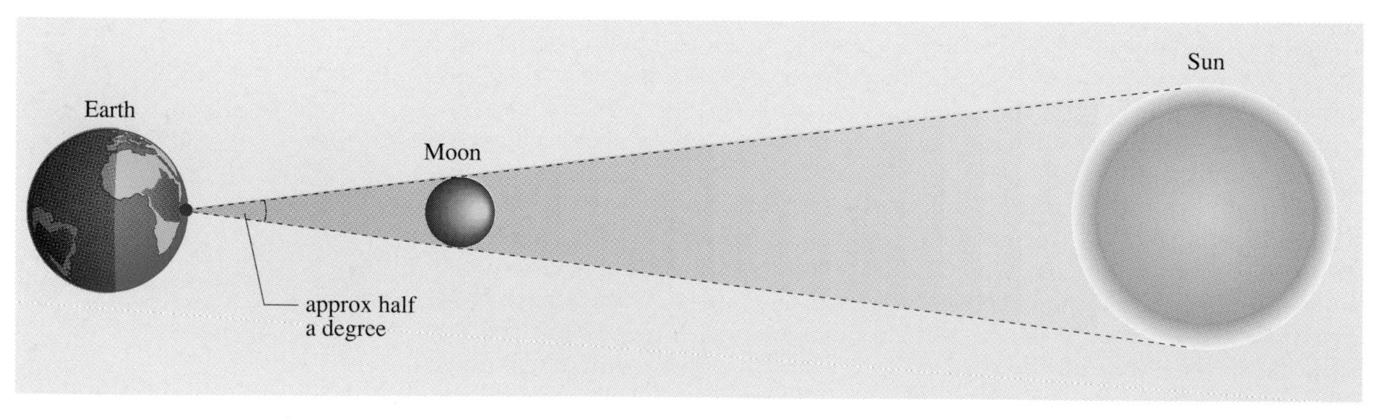

NOT TO SCALE!

The Sun and the Moon have about the same angular diameter because, though the Sun's linear diameter is about 400 times that of the Moon, it is also about 400 times farther away. NB Not to scale!

The images in this book are arranged to tell a story, so it makes sense to go through them in the order in which they are presented. Wherever possible, the images are presented with north at the top, and at the end of the book there are maps showing where the objects are located in the sky. Most of the images have been obtained at visible wavelengths, but we always say when this is not the case. The non-visible images have been produced by converting the output from the radiation detector into various shades of grey, representing different brightnesses at a single wavelength. Sometimes these shades are translated into a range of colours, called *false colours*. False colours are also used to represent *different* non-visible wavelengths, just like the colours at visible wavelengths. They can be used in other ways too, as you will see.

At the end of the book is a glossary of some of the less common words that we've used in our story.

So, that's the end of the preliminaries: it's time for your journey to begin.

Part 1

The stars and the interstellar medium

The Sun is an average sort of star. Like the other stars, it was born from the interstellar medium – the dust and gas that pervade the vast spaces between the stars. This medium is dominated by the lightest element, hydrogen, present as hydrogen gas. It is a medium densest where it is coolest, in what are called *dense clouds,* though they are far less dense than our atmosphere even at high altitudes. It was in such a dense cloud that the Sun was born, and starbirth continues in such clouds today.

Stars are born in clusters of up to a few hundred, the remnants of the dense cloud then becoming dispersed. These clusters are called open clusters: many of them break up to leave isolated stars like the Sun, or compact groups of a few stars, often just two – a binary star.

A new-born star consists mainly of hydrogen; it shines because of nuclear reactions in a fiercely hot core, which convert the hydrogen into the next heaviest element, helium. When this core conversion is taking place, the star is said to be in its main sequence phase. This phase accounts for nearly all of a star's lifetime, its length depending strongly on the mass of the star: for a star with the mass of the Sun (one solar mass) the main sequence phase lasts about 10 thousand million years, whereas for a star of half a solar mass it lasts an enormous 200 thousand million years, but only 15 million years for a star of 15 solar masses.

Stellar mass is important in other ways too. For example, the greater the mass of a main sequence star the larger it is, and the greater its interior and surface temperatures. The surface temperature determines the visual tint of the star. The coolest stars appear orange-white, then comes yellowish-white, then white, then, with the hottest stars, bluish-white. The Sun is yellowish-white.

After the hydrogen in its core is used up, a star's evolution is comparatively rapid. The star swells and its surface cools, details depending on the mass of the star. Stars with masses up to about 8 times that of the Sun become red giants, with diameters 10 to 100 times the main sequence diameter, and with an orange or yellowish tint. New nuclear fuels are consumed, with heavier and heavier elements being formed. The next major event for such a star is the shedding of a good deal of its mass in the form of an expanding gas shell called a planetary nebula. The stellar remnant shrinks to become a tiny (Earth-sized) white dwarf. Initially the white dwarf is very hot, but it is devoid of nuclear fuel and so it cools, becomes less bright and eventually vanishes from view.

Stars with masses greater than about eight times that of the Sun end their main sequence lives by expanding and cooling to become supergiants. After a relatively short time they then explode, in an extraordinarily bright phenomenon called a supernova. The expanding shell is impressive, but ultimately dims. If any of the star remains, it will probably be as a tiny ultradense neutron star, in which a solar mass or so is packed into a volume only about 10 kilometres across. Some neutron stars are observed to emit astonishingly regular pulses of radio waves – the star is then called a pulsar. Another possibility is that the remnant becomes a black hole, into which matter, light, everything, vanishes from view.

If the members of a binary star are very close together, they can influence each other's evolution, mainly by transferring mass from one to the other. One possible outcome is that the star receiving mass suffers an enormous outburst, called a nova.

The matter expelled by the stars at the ends of their lives replenishes the interstellar medium. New dense clouds form, leading to new generations of stars. There is thus a cosmic cycle. The cosmic cycle is not closed: some matter remains locked up in dead, cold stellar remnants, such as cold white dwarfs, and new matter arrives from beyond the Milky Way. There is also a gradual change in chemical composition because of nuclear reactions in stars, turning hydrogen into heavier elements. However, the cosmos is still young, in that it is still dominated by hydrogen, and in that the cosmic cycle, much as it is operating today, has been in existence since not so very long after the beginning.

The first set of plates in Part 1 are of the Sun – our star. Subsequent images are arranged to follow the cosmic cycle, starting with the interstellar medium and with starbirth in dense clouds. We then look at the stars themselves, ending with star death and the return to the interstellar medium of some of the products of their death throes.

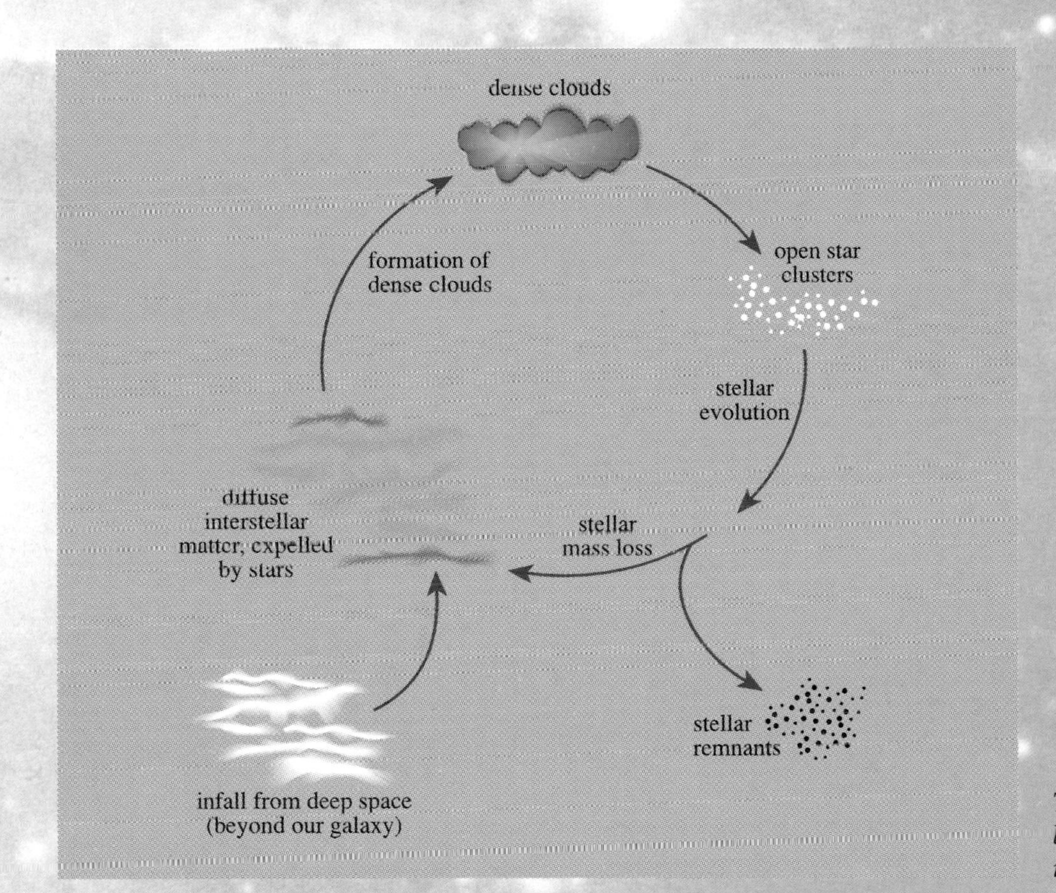

The cosmic cycle, operating between the stars and the interstellar medium.

1.1 The solar corona during a total solar eclipse

A total solar eclipse occurs when the Moon passes between the Earth and the Sun, obscuring the bright solar surface. This eclipse occurred on 16 February 1980. The black disc in the middle of this image is the Moon. The colourful pattern around it is 400 times more distant, 150 million kilometres away, and is the extended outer atmosphere of the Sun – the solar corona – very hot, with temperatures of millions of degrees Celsius (°C), but so thin that it is not normally visible. Total eclipses can last for up to 7 minutes. They are spaced, on average, by a little over a year, but each one is visible from only a narrow path on the Earth's surface.

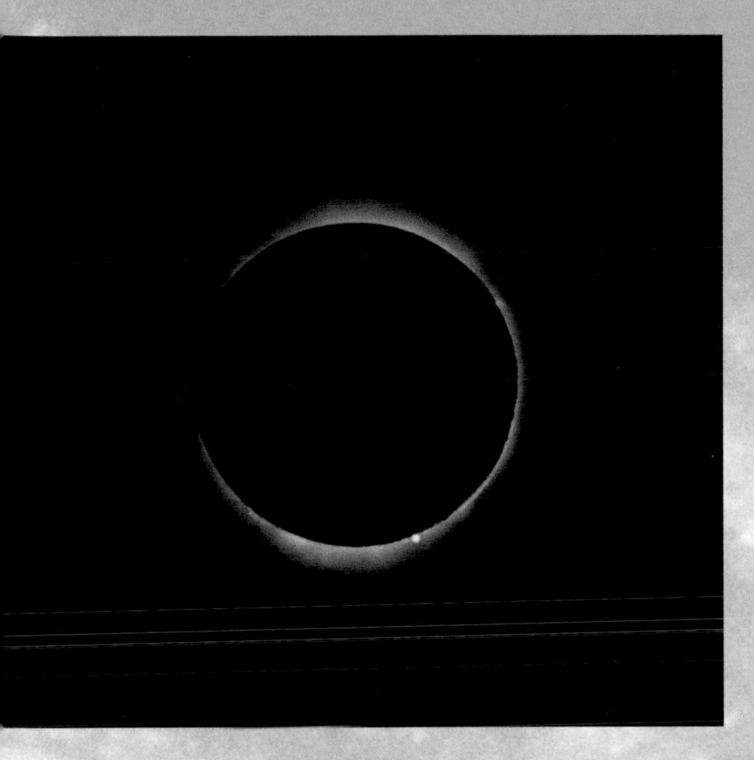

1.2 The solar chromosphere during a total solar eclipse
The total solar eclipse of 10 July 1972 is caught here at a moment when the Moon has only just covered the brilliant surface of the Sun, and is not yet centred on it. Therefore, a segment of the lowest part of the Sun's atmosphere is visible as a red arc. This is the chromosphere. It is uneven because of flame-like prominences that protrude into the corona (Plate 1.1). The corona is not visible because of the short exposure time used to obtain this image.

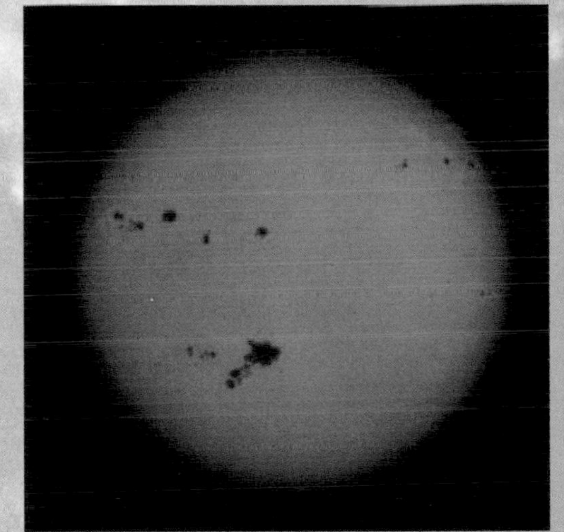

1.3 The solar photosphere, at high and low solar activity
The brilliant surface of the Sun, the solar photosphere, is not a solid surface but hot, glowing gas at a temperature of about 5500 °C. It is about 500 kilometres thick, a tiny fraction of the solar diameter of 1.4 million kilometres. The two images differ in the number of dark sunspots visible. These are slightly cooler regions of the photosphere that look dark only through contrast. They are transitory features, and are one indication of the level of solar activity, there being more spots at times of high activity than at times of low activity. Solar activity follows a rough cycle, about 11 years long. The unspotted image was obtained on 16 June 1986, near a time of minimum activity, whereas the spotted image was obtained on 13 June 1989, near a time of maximum activity.
WARNING Never look at the Sun for more than an instant with the unaided eye, and not even for an instant with binoculars or a telescope, otherwise you will suffer permanent eye damage, even blindness.

1.4 Various types of spectrum

The nature of the Sun, or any other body, can be investigated by spreading the light it emits into a display of brightness versus wavelength. Such a display is called a spectrum. The three main kinds are shown here. A hot, dense material emits light across a continuous range of wavelengths, giving a *continuous spectrum.* If light from such a source passes through a gas then some of the wavelengths are absorbed by the gas, and are then emitted in all directions. Thus, if we look towards the gas in a direction other than towards the source, we see the wavelengths emitted by the gas as a set of bright lines on a dark background. This is an *emission line spectrum.* If we look towards the gas in the direction of the source the continuous spectrum displays *dark* lines at the absorbed wavelengths. This is an *absorption line spectrum.*

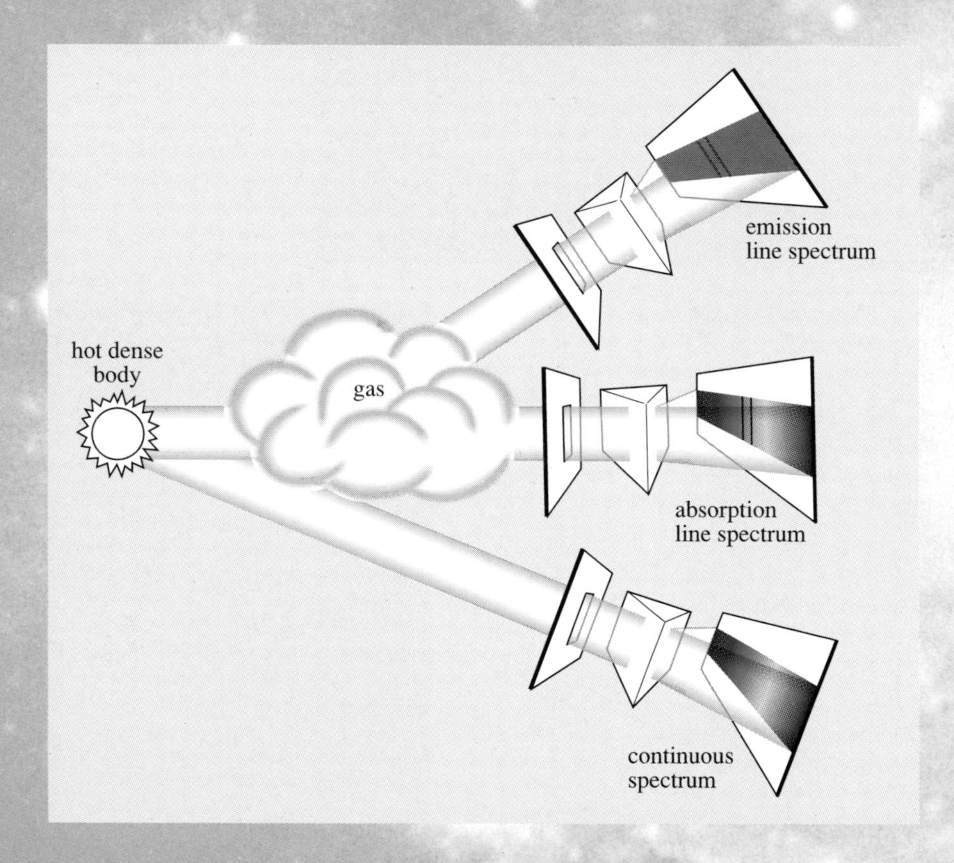

1.5 Emission line spectra from various chemical elements
The particular set of wavelengths at which the lines in an emission spectrum occur depends on the nature of the atoms in the gas. Thus, from an emission line spectrum we can deduce what chemical elements are present in the gas. Here, from top to bottom, the elements are: hydrogen, mercury, helium, neon, copper, zinc and iron. Likewise, we can do the identification from an absorption line spectrum, such as that of the Sun in the next plate.

1.6 The solar absorption line spectrum
The hidden interior of the Sun gives a continuous spectrum, but what we actually see is the radiation after it has been absorbed and re-emitted by the photosphere and then passed through the solar atmosphere. The result is a richly detailed absorption line spectrum. This visible spectrum alone contains thousands of lines, and there are yet more at infrared and ultraviolet wavelengths. From these lines we learn, for example, that the solar photosphere and atmosphere consist largely of the lightest chemical element, hydrogen – a minor constituent of the Earth.

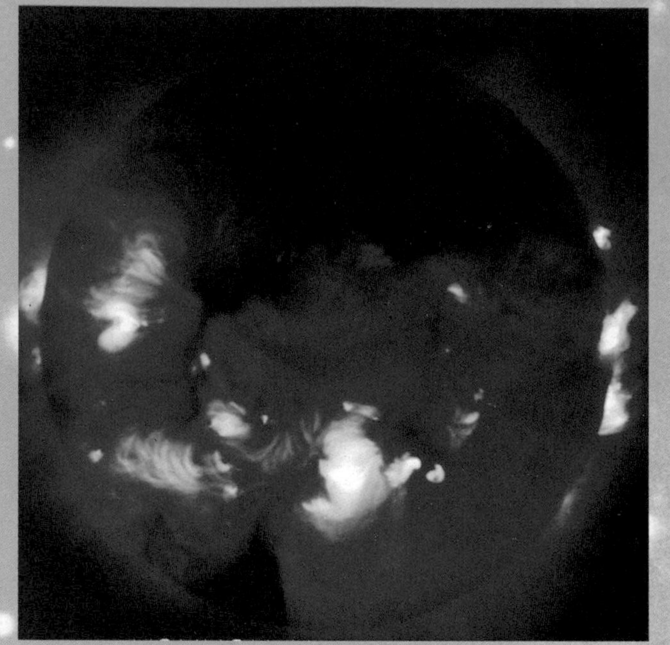

(a)

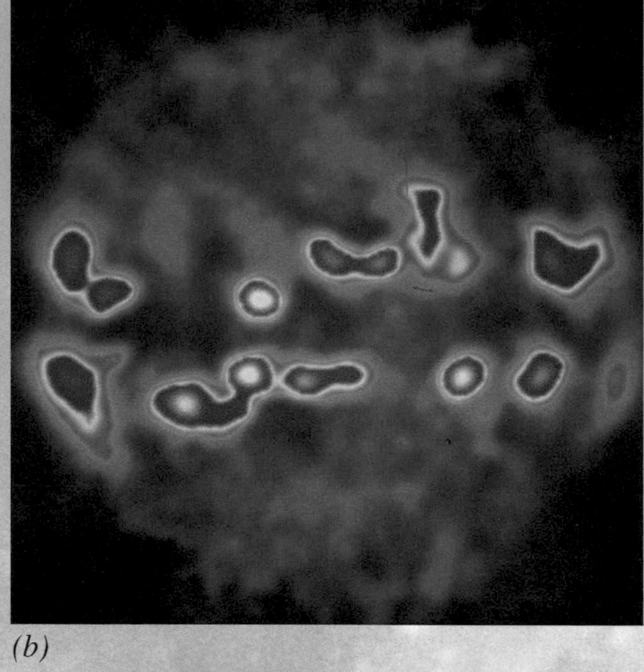

(b)

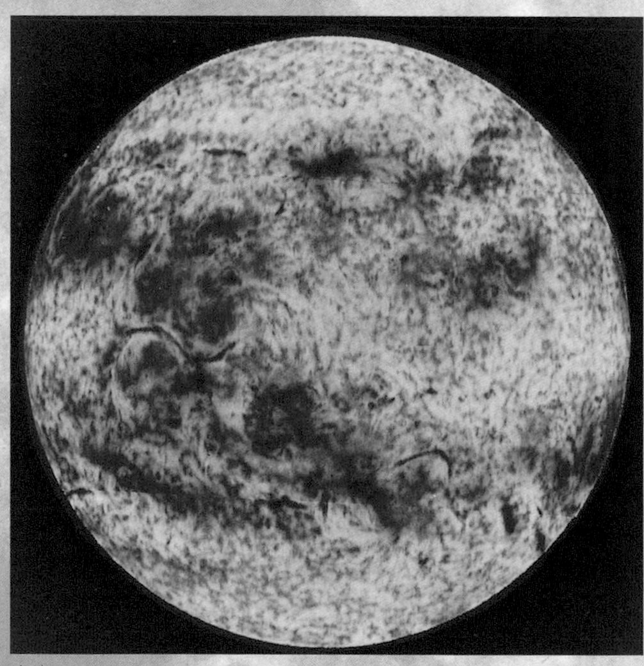

(c)

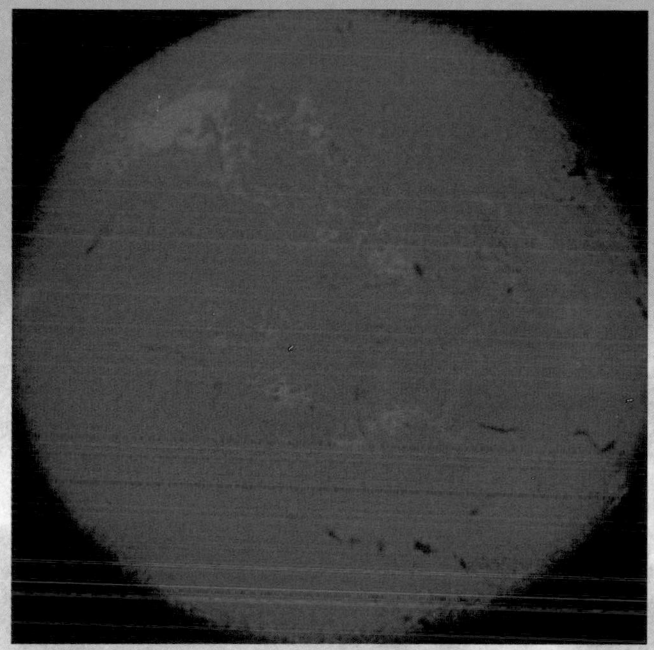
(d)

(e)

1.7 The Sun at various wavelengths

Though much of the Sun's radiation is at visible wavelengths, there is a significant proportion at other wavelengths. (a) This X-ray image was obtained by a telescope in orbit above the Earth's atmosphere (which is opaque to X-rays). X-rays are generated in energetic processes, and here they are produced largely by the enormous temperatures in the corona, typically 1.6 million degrees Celsius. A lot about coronal structure is learned from these images. (b) The corona can also be studied via radio wave images like this. (c) Just beyond the visible spectrum, in the near infrared, a wavelength emitted by helium atoms reveals filamentary structures in the higher parts of the chromosphere. (d) This image is at a wavelength in the red part of the visible spectrum, a wavelength called H-alpha, emitted by hydrogen atoms. This emission is particularly strong from the chromosphere, so we learn about chromospheric structure from H-alpha images. (e) When the whole visible spectrum is used we get an image of the photosphere, and its sunspots.

1.8 A large solar prominence
Solar prominences extend up from the chromosphere. They are huge clouds of relatively cool gas (10 to 20 thousand degrees) amidst the far hotter but far less dense corona. Prominences are lifted by magnetic fields, and the dramatic eruptive kind, as here, usually last less than a day before sinking back down, or dissipating outwards. The more quiescent kind can last for weeks or months. Like sunspots (Plate 1.3) they are an indication of the level of solar activity, being more common at times of high activity. This image was obtained at an ultraviolet wavelength emitted by helium atoms that have each lost one electron. The photosphere is not particularly bright at ultraviolet wavelengths, enabling prominences to be seen readily.

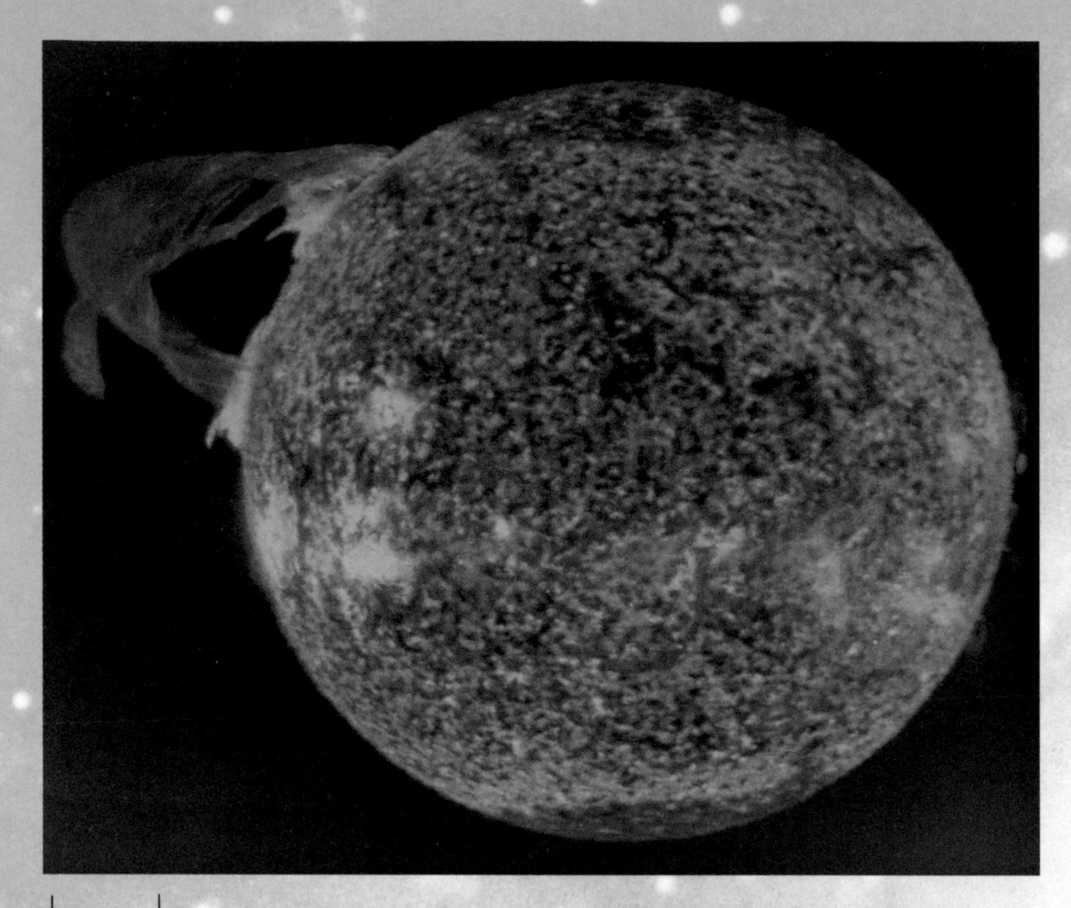

200 000 km

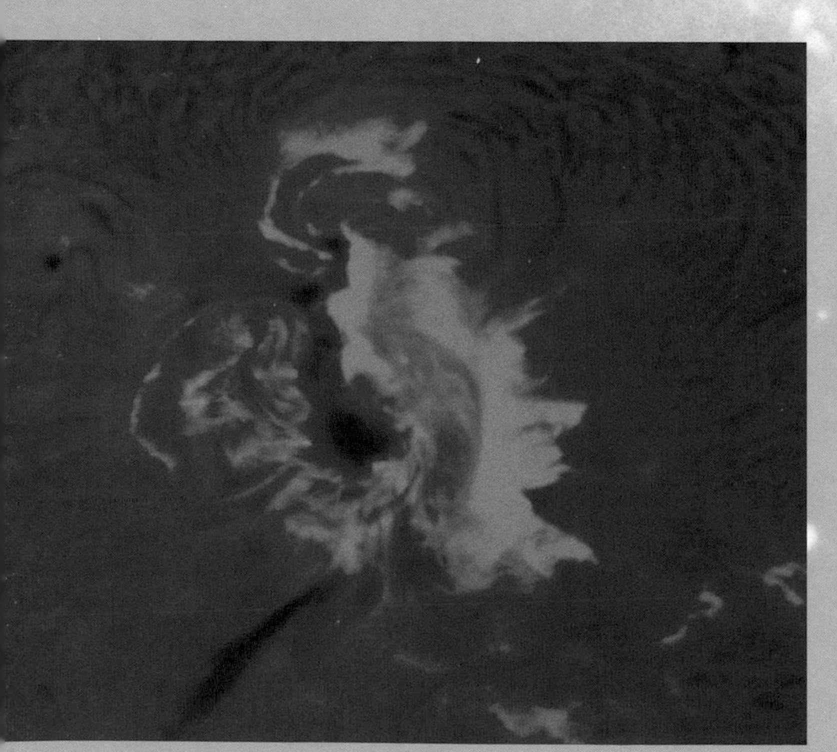

50 000 km

1.9 A solar flare
A solar flare is a sudden brightening of the corona and chromosphere because of a rapid release of stored energy (probably magnetic in origin). A flare lasts no more than an hour. Like sunspots (Plate 1.3) and prominences (Plate 1.8) they are an indicator of solar activity, being more common at times of high activity. Flares can be observed at many wavelengths, including at H-alpha, as in this image, where the flare appears bright against the photosphere. Flares also cause an increase in the s*olar wind* – a gusty stream of particles moving rapidly away from the Sun, carrying magnetic fields with it. The wind consists mainly of the separated constituents of hydrogen atoms, namely protons and electrons. The X-rays from a powerful flare can disrupt radio communications on Earth, and the increased solar wind can give rise to spectacular auroral displays.

(a)

(b)

1.10 Auroras seen from the ground and from space
Auroras are spectacular displays in the Earth's upper atmosphere occurring over the polar regions: (a) as seen from ground level in northern Sweden, and (b) between Australia and Antarctica, as seen from the Space Shuttle. Auroras are caused by the solar wind, and are therefore more common and more spectacular at times of high solar activity. The protons and the electrons in the wind are electrically charged, so when they encounter the Earth's magnetic field (at speeds up to 1000 kilometres per second), they are swept towards the polar regions, where they strike the Earth's upper atmosphere. Gases in the atmosphere absorb the energy from these particles, and re-emit some of it as light, to produce the aurora.

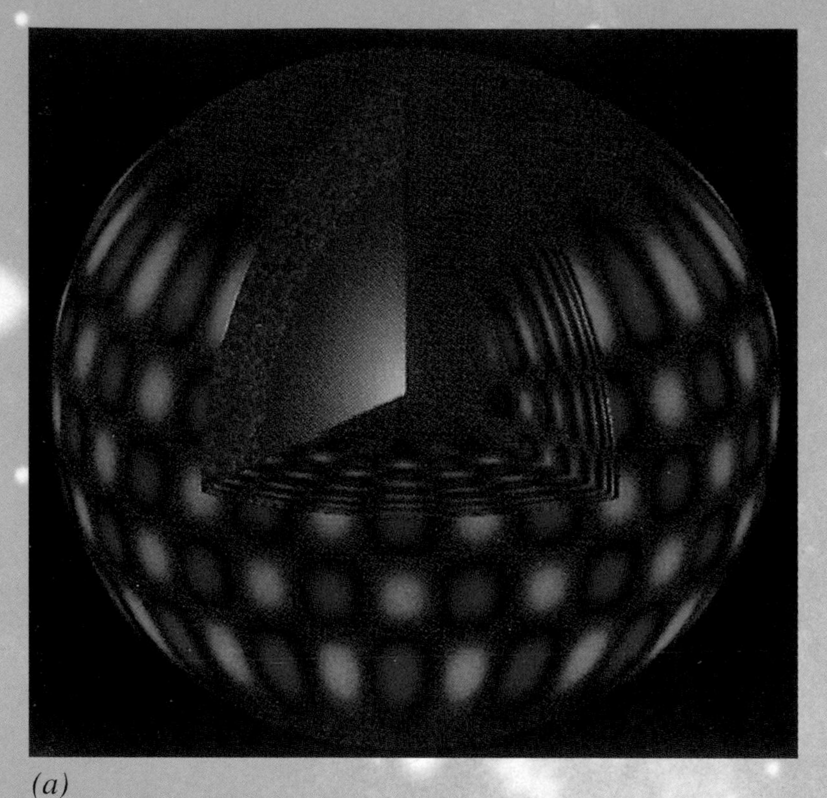

(a)

1.11 Solar oscillations and solar rotation
(a) A remarkable discovery in the 1960s and 1970s was of a great variety of patterns of vertical oscillations covering the whole solar photosphere, such as the one shown here, frozen at a particular moment. The overall motion is a combination of many patterns – this is just one of them. The oscillations penetrate deep down, and so help us to understand the Sun's interior, rather as earthquakes help us to understand the Earth's interior. A typical oscillation takes about five minutes to go through a complete up-and-down cycle, which repeats over and over again. (b) Solar oscillations have helped us to establish the rotation periods at different depths in the Sun. Though the central volume is fluid, it all rotates with a period of 27 days. By contrast, the photosphere has long been known by direct observation to have a rotation period of 25 days at the solar equator, increasing to 36 days at the poles, and we now know that these periods persist to a considerable depth. The various periods are colour coded here.

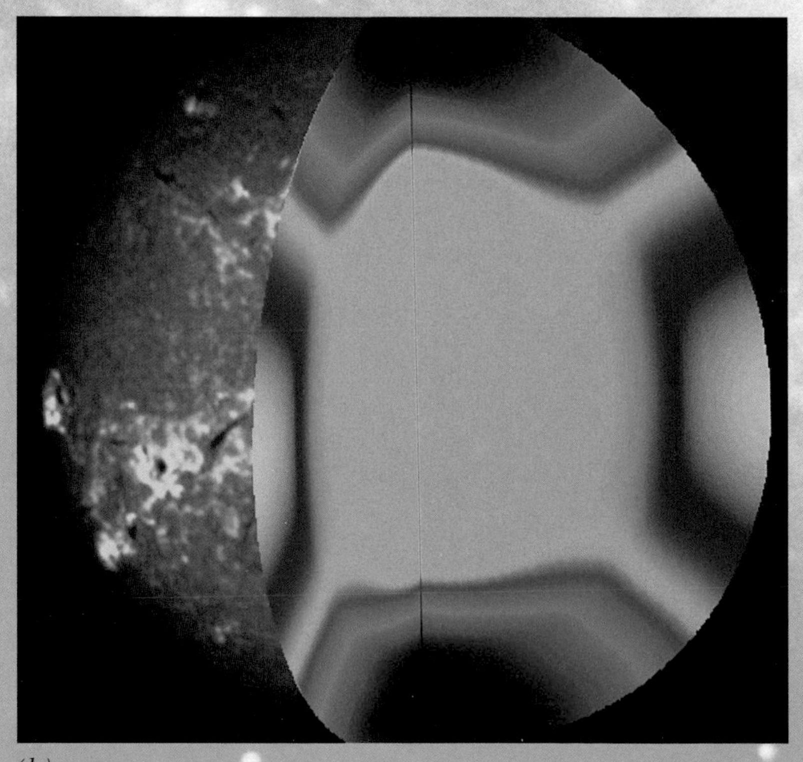

(b)

1.12 The solar interior

Though the interior of the Sun is hidden from view, we can infer much of what is going on by making full use of a great variety of solar observations. This model shows some of the widely agreed features. The temperature and density increase rapidly with depth, reaching 10 million degrees Celsius and ten times the density of lead in a central core that is hot enough and dense enough to sustain the nuclear reactions that are the Sun's energy source. These reactions convert hydrogen into helium. Outside the core the energy diffuses outwards in the form of radiation, until the convective zone is reached, where the energy travels outwards by convection, just as in a pan of heated water. It takes about a million years for the energy to travel from the core to the photosphere.

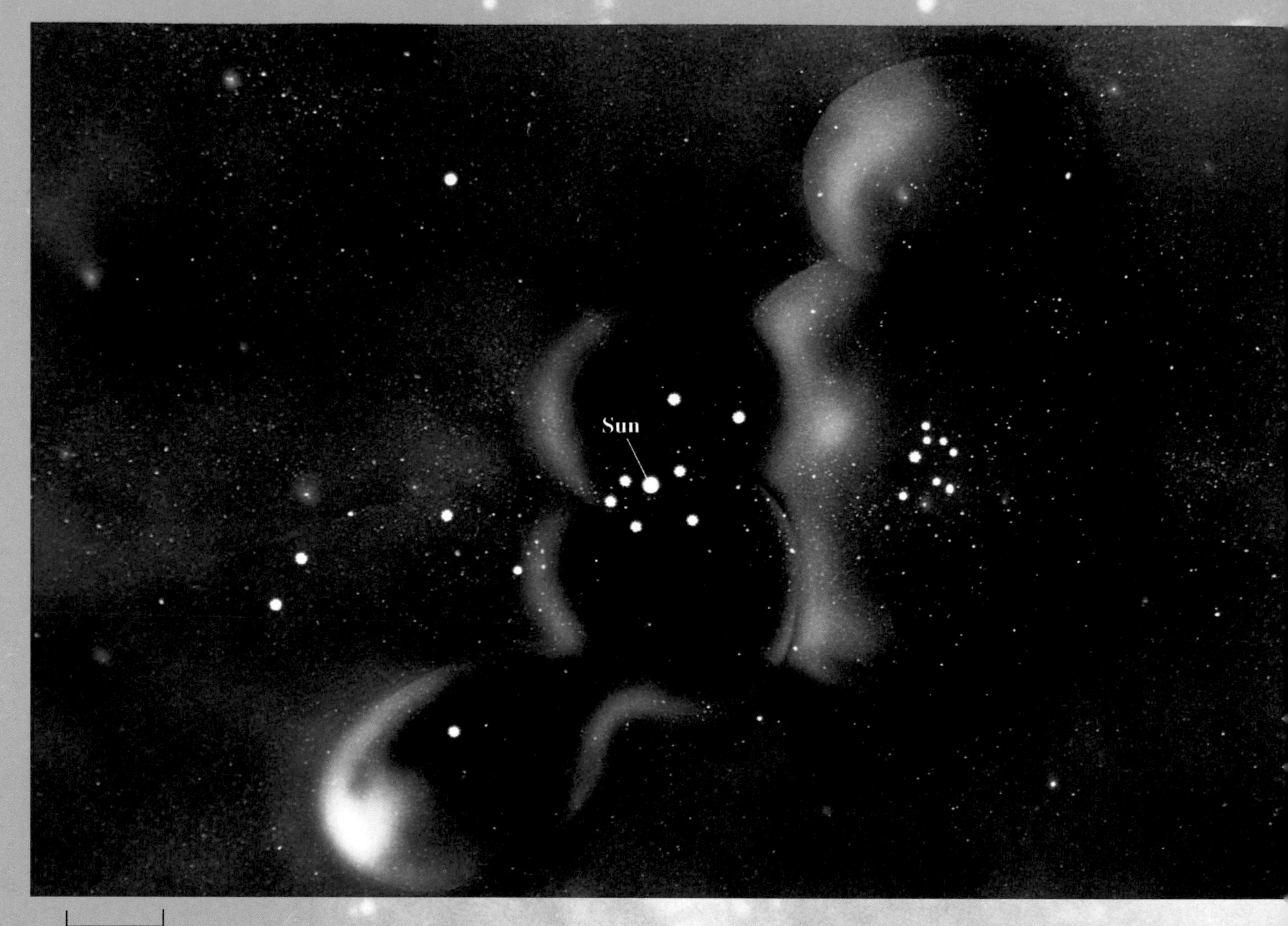

1.13 The local bubble in the interstellar medium

When we travel away from the Sun, to well beyond the planets, we enter the enormous spaces between the stars. This space is not empty but contains the gas and dust that constitute the interstellar medium, or ISM. It is richly varied, some parts hot and rarified, other parts cool and dense, though even at its densest it is much thinner than the Earth's atmosphere. In the neighbourhood of the Sun the ISM is hot, perhaps because of a star that exploded at least a hundred thousand years ago, but we don't see it glowing across the sky because it is extremely rarified, even by ISM standards. Our interstellar neighbourhood is called the local bubble, and it is irregularly shaped. Its high temperature means that the emitted radiation is concentrated at short wavelengths, and so the local bubble has been studied by sensitive instruments at ultraviolet wavelengths (above our atmosphere).

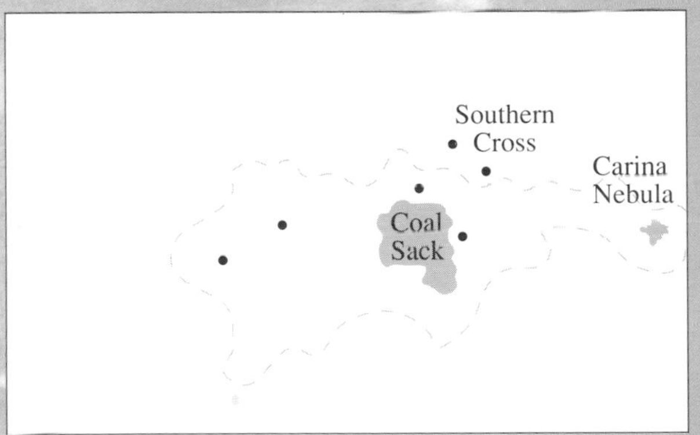

1.14 A panorama of the southern skies
Several different types of region in the interstellar medium are visible here. The Coal Sack is not a star-free tunnel, but a cool dense cloud, the dust in it obscuring the light from the stars beyond. It is about 500 light years away, and about 50 light years across. The rosy Carina Nebula is a glowing gas cloud about 9000 light years away, about 200 light years across ('nebula' means 'cloud'). It is lit by young stars embedded in it. Near the Coal Sack is the famous Southern Cross, the different star colours having been exaggerated in this image.

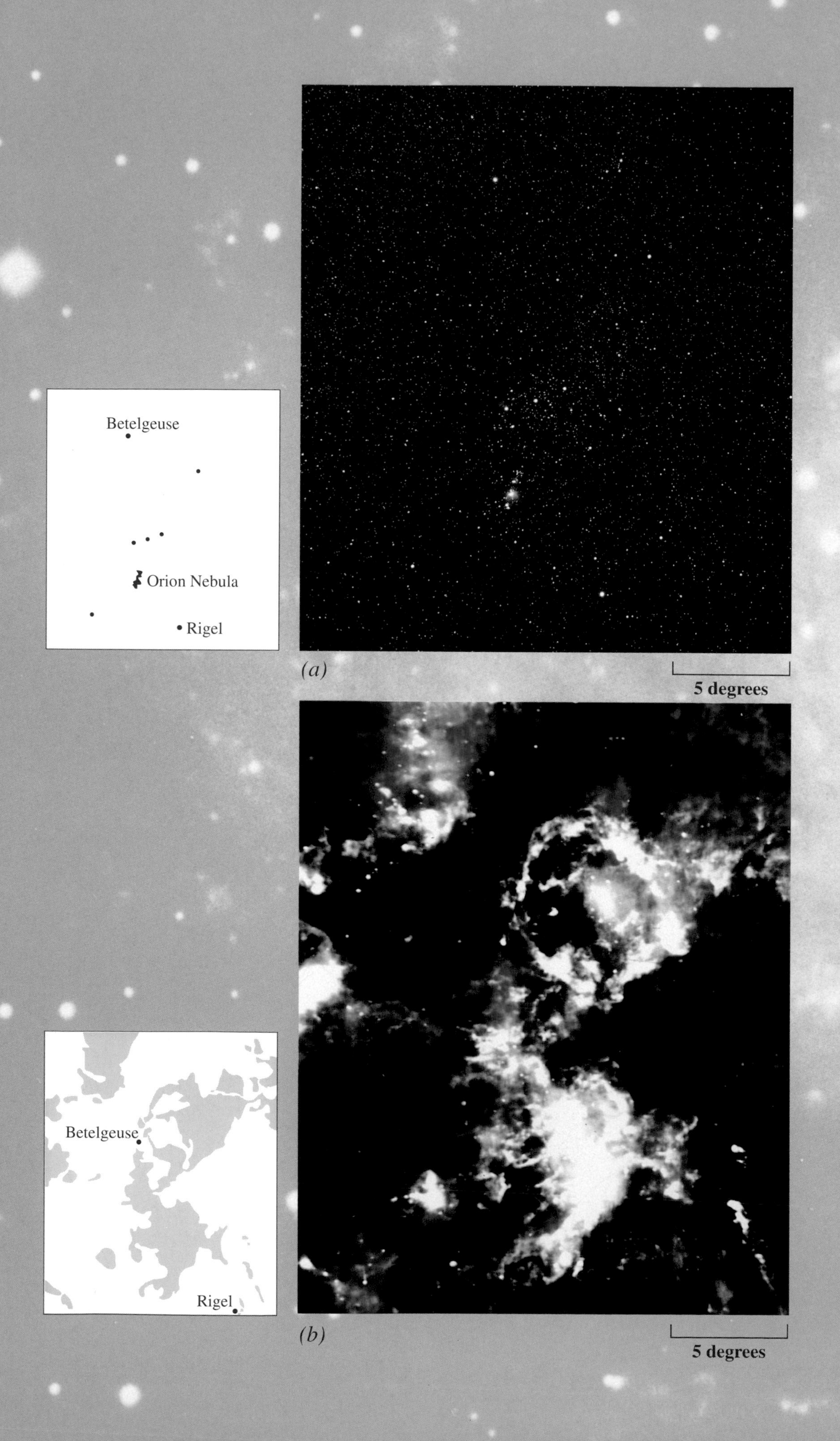
Betelgeuse
Orion Nebula
Rigel
(a)
5 degrees
Betelgeuse
Rigel
(b)
5 degrees

5 degrees

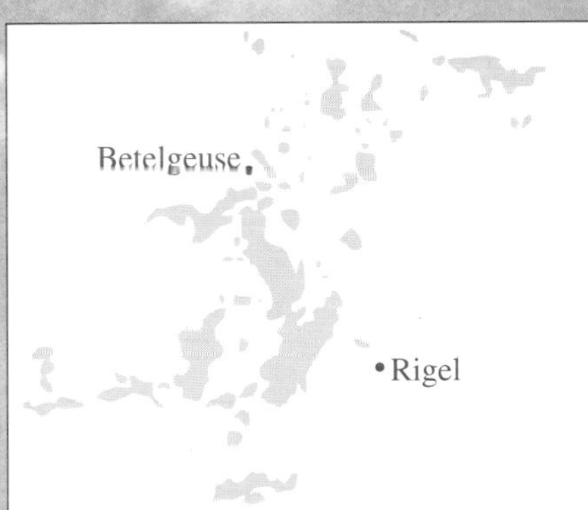

1.15 The Orion region at various wavelengths
The famous constellation Orion is an area of sky rich in stars and interstellar matter. (a) At visible wavelengths the region is dominated by stars. Betelgeuse and Rigel are supergiant stars, several hundred light years away. The Orion Nebula is also visible – a star-lit glowing gas cloud, 1500 light years away, 13 light years across. (b) This is much the same area of sky as in (a), but now at infrared wavelengths. Cool interstellar dust is now prominent – the redder the cooler (false colour). The whole area is seen to be pervaded by clouds of dust, and, by implication, the associated gas. Most of these clouds are where young stars are being born, but to the right of Betelgeuse is a ring-like object that is a shell of matter swept up by the shock wave created by an exploded star (a supernova). Most of the objects here are 1000 to 2000 light years away. (c) This is much the same region as in (b), but now at a radio wavelength emitted by carbon monoxide molecules. These molecules pervade the cool regions of the interstellar medium and are relatively easy to detect. They gives a reliable map of where the gas (mainly hydrogen) lies. The density of the gas is indicated by the colour: black where it is undetectable, violet where it is least dense, then through the colours of the rainbow to red where it is most dense.

1.16 The Orion Nebula: star formation in a dense cloud
This is a close-up of the glowing gas just visible in Plate 1.15a. The gas, mainly hydrogen, is made to glow, in the main, by four very bright, very massive stars, of up to about 50 solar masses, which are located in the centre of the brightest region. These stars are called The Trapezium, and are part of a very young cluster of a few hundred stars, born less than a million years ago. The dense cloud that gave birth to this cluster is apparent through the obscuration caused by the dust in it. The glowing gas is just on our side of the cloud, and is material left over after star formation.

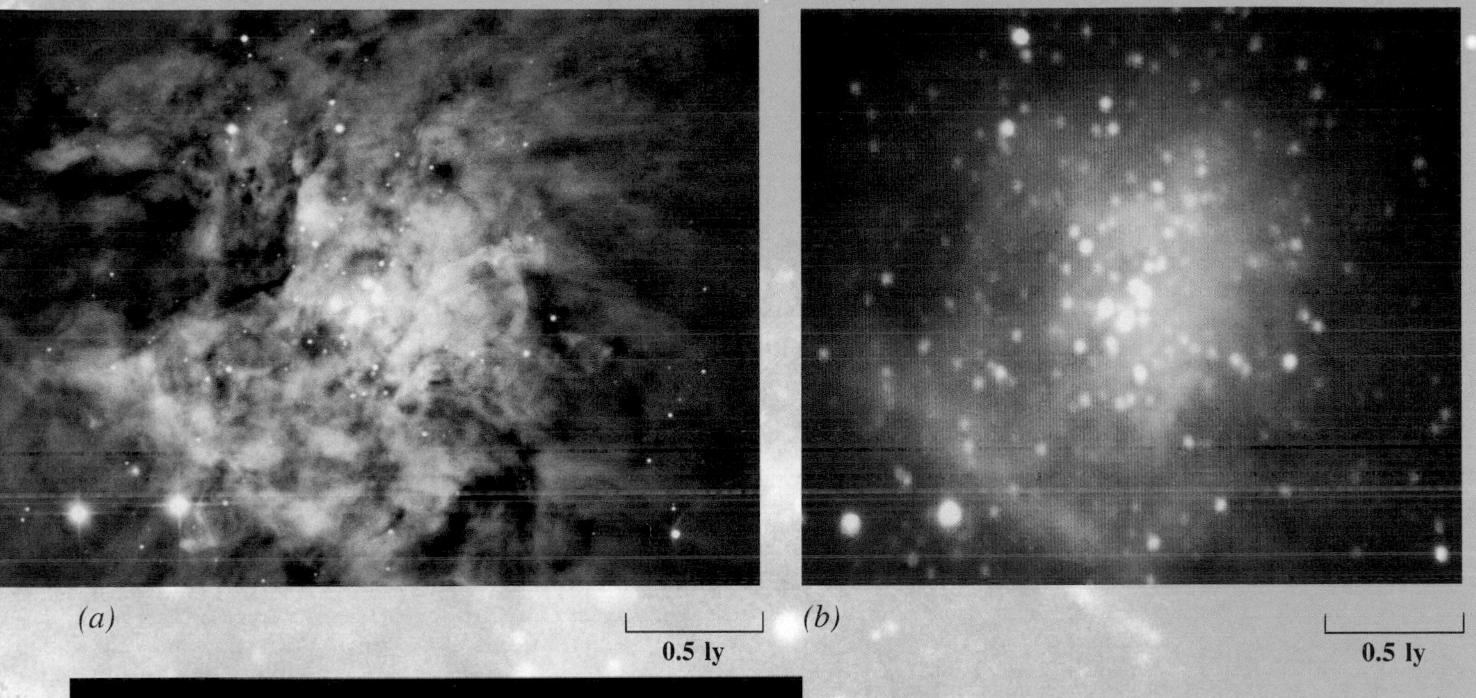

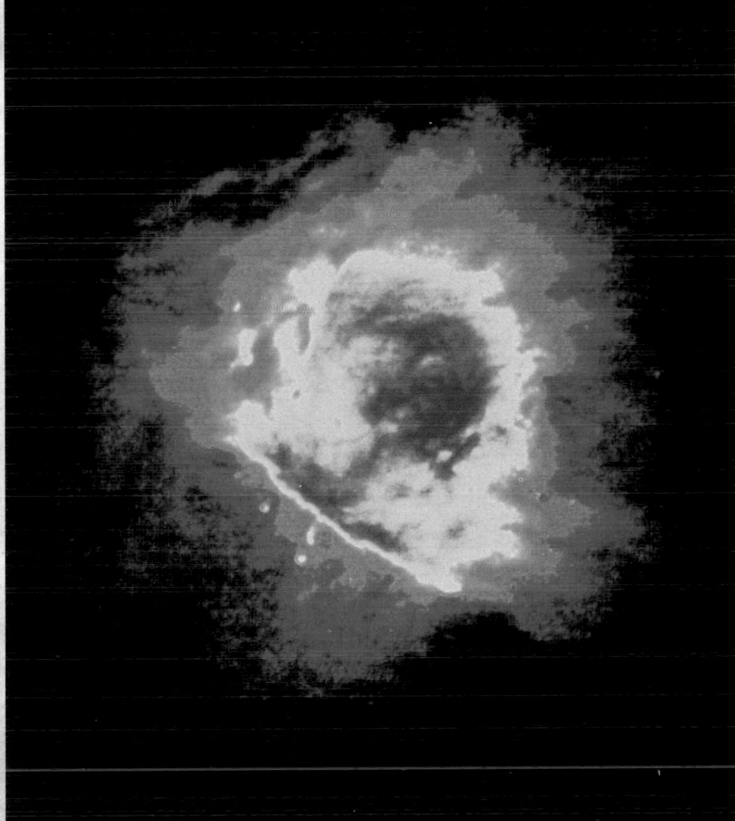

1.17 The Trapezium region
(a) The four Trapezium stars in the Orion Nebula (Plate 1.16) are near the centre of this image. Some of the other stars visible were also born near the surface of the dense cloud, but others are foreground stars. (b) At infrared wavelengths the dust is less obscuring, and so we can see not only the Trapezium stars but some way into the dense cloud, where young stars are forming, or have recently formed. (c) At radio wavelengths dust obscuration is negligible, and so too is stellar radiation scattered by the dust. Therefore, by mapping the radio waves that the gas emits, we can see the complete distribution of the gas in and around the dense cloud. Here we see gas that has been heated by star formation.

1.18 Interstellar matter moulded by star formation

(a) This image is of a small region near The Trapezium (Plate 1.17). Many details are visible, showing the dense cloud being consumed and racked by star formation. For example, the V-shaped feature near the edge is where an invisible outflow from a forming star is colliding with other interstellar matter: in star formation, though the net infall must be inwards from far less dense material, some of this matter flows out again. Two circumstellar discs around young stars are also visible, as small orange discs, one near the right arm of the V-shaped feature.

(b) In this radio image of the object IRS5 (about 520 light years away, within the dense cloud L1551), the outflow from a young star is clearly visible. It consists of two oppositely directed lobes – the one coloured blue is flowing towards us and the one coloured red is flowing away. These lobes together constitute bipolar outflow. The radio waves are emitted by carbon monoxide molecules in the gas swept up by the material streaming away from the star. The swept-up gas consists mainly of hydrogen molecules. The star itself is at the position of the cross. A ring or disc of dense cool gas and dust is thought to separate the two lobes, and would be presented nearly edgewise to us. From such rings or discs planetary systems could form. Discs and the associated lobes of bipolar outflow are common features of young stars.

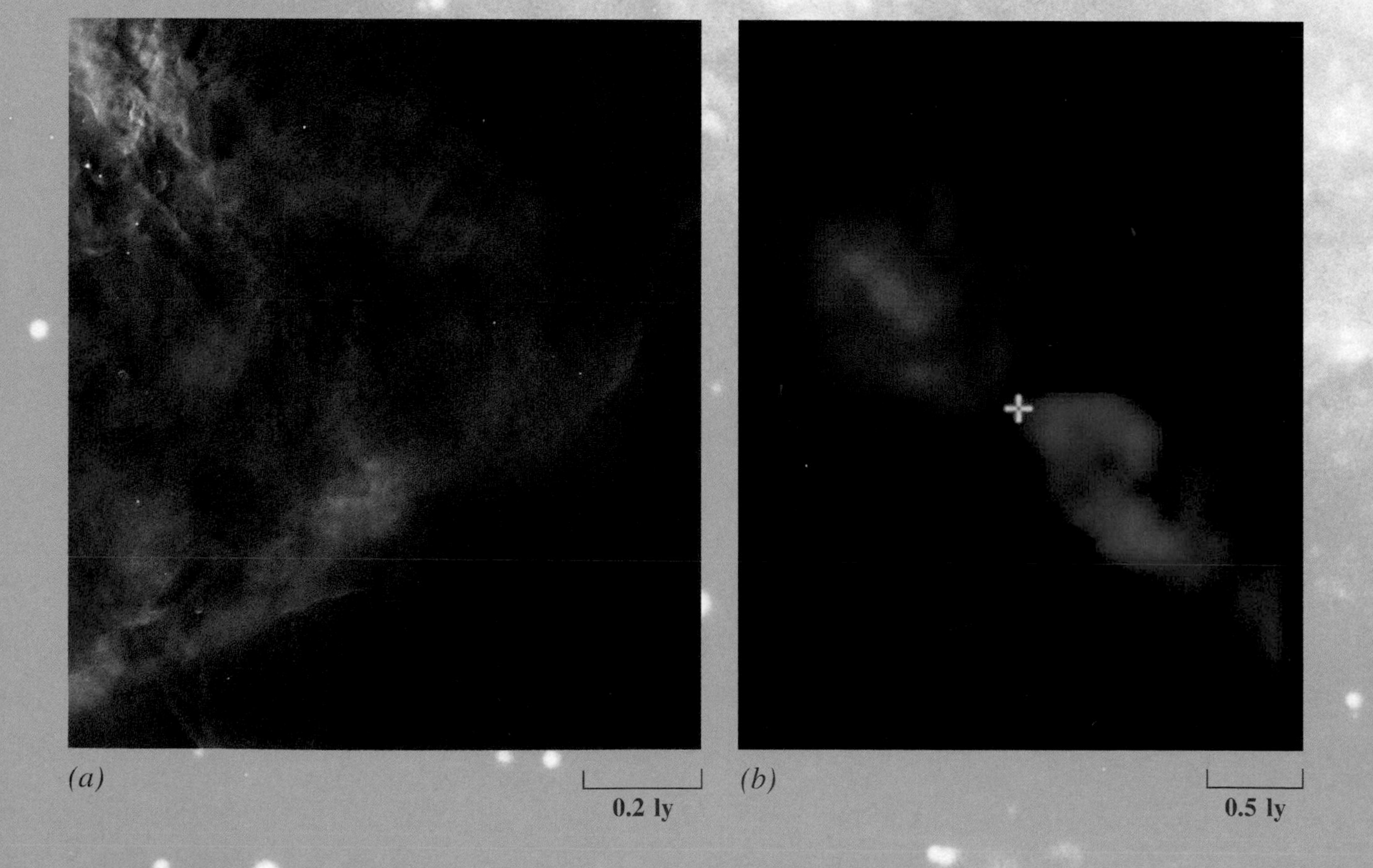

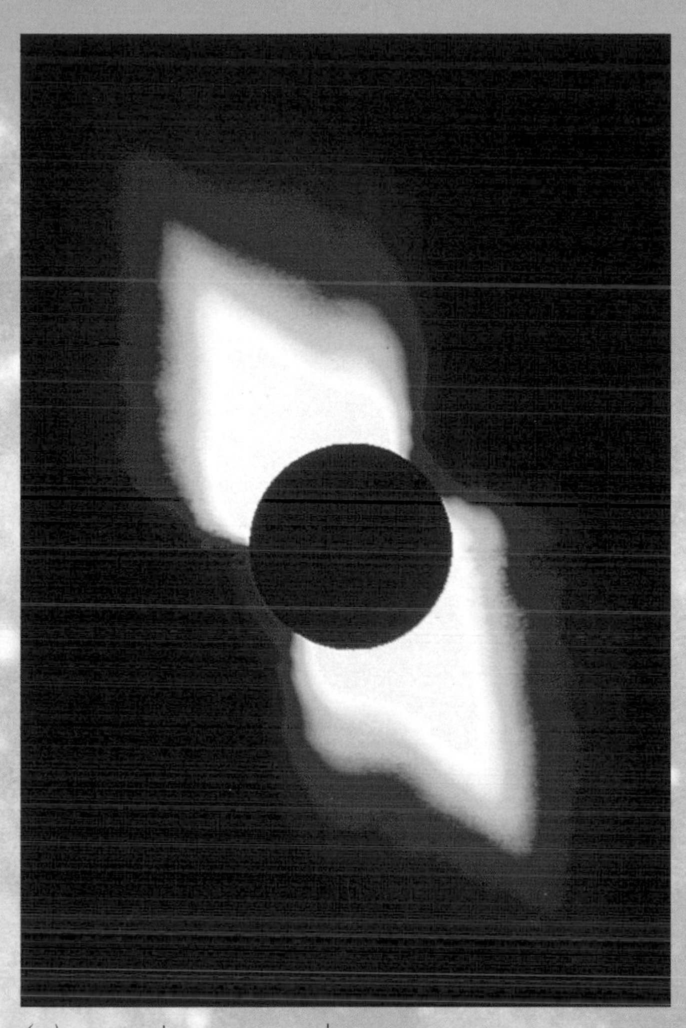

1.19 The disc around the star Beta Pictoris
(a) This near-infrared image is centred on the young star Beta Pictoris, which is only a few hundred million years old. The black circular patch, much larger than the star, prevents the radiation from the star from swamping the image of the much feebler radiation from a disc of dust around the star. Using the infrared wavelengths emitted by the warm dust further enhances the contrast with the star. The disc is presented to us almost edge-on. From such a disc a planetary system could be forming, though our own planetary system would be hidden behind the patch. Beta Pictoris is 53 light years away. (b) Hot gas has been detected at ultraviolet wavelengths in the inner regions of the Beta Pictoris disc, as shown obliquely in this artist's impression. Much of this gas is spiralling in to the star, to complete its formation. The Earth's orbit would have a radius of about 20% of that of the gas disc here.

1.20 The Trifid Nebula: star formation in a dense cloud
There are many dense clouds exhibiting star formation, many of them (Plate 1.16) as impressive at visible wavelengths as this one, the Trifid Nebula, 3500 light years away The rosy colour is hydrogen gas, made to glow by bright young stars visible near where the three dust lanes meet. The blue light is starlight scattered by dust.

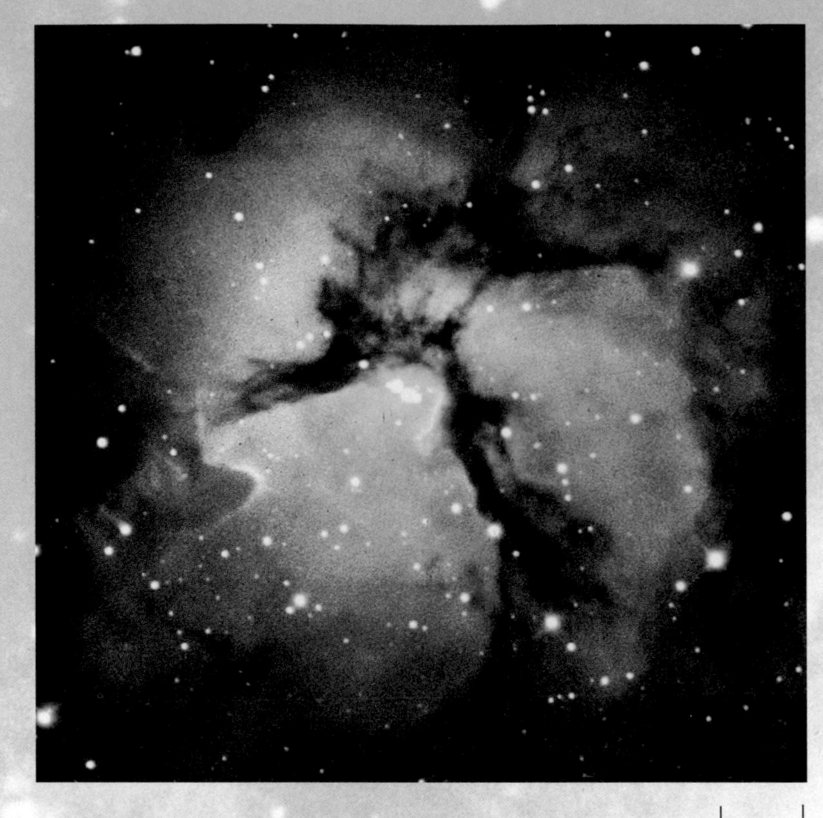

1.21 In and around Rho Ophiuchi: star formation in dense clouds
(a) The warm dust in an extensive star-forming region is revealed in this infrared image. The dust is warmed by numerous young stars that have formed within dense clouds, of which the dust is a minor component. The region is about 500 light years away. (b) At visible wavelengths, we see the star Rho Ophiuchi at the top, embedded in the remnants of the dense cloud that gave it birth. Other regions of star formation are also visible.

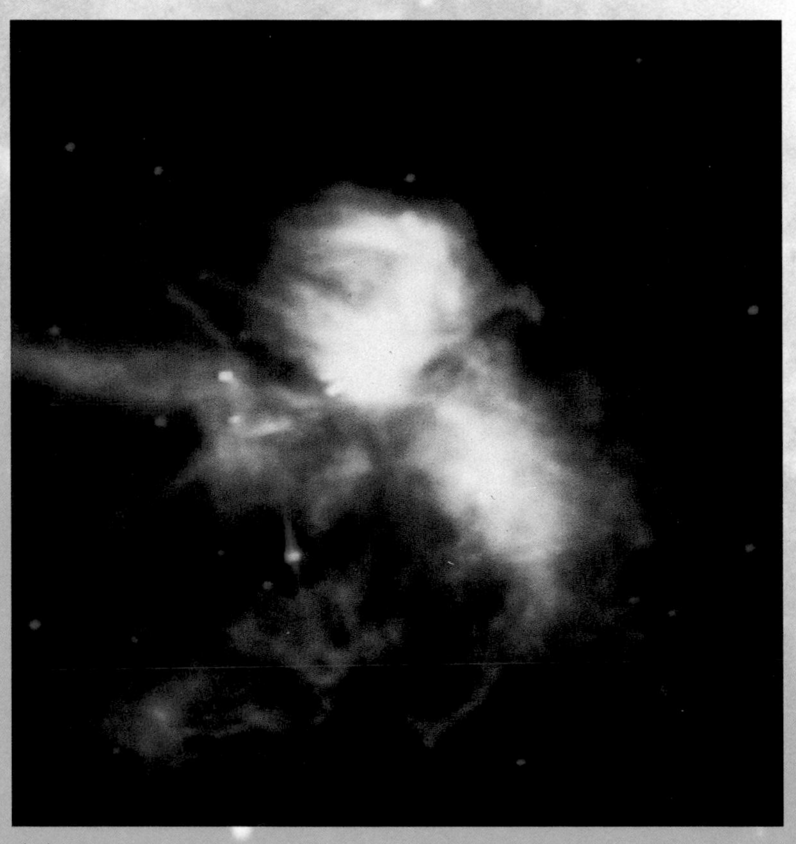

(a)

(b)
2 ly

(a)

1.22 Star clusters, young and old
In dense clouds stars form in open clusters of several hundred. Some open clusters break up after many millions of years, or lose stars, to yield isolated stars like our Sun. (a) This is a very young open cluster: only about 5 million years old, and the lit-up dense cloud that gave it birth is still very much in evidence in this visible image. It is 2400 light years away. The cluster has no name, but carries the catalogue number NGC 2264.

(b)

(c)

(b) The open cluster The Pleiades is readily visible to the unaided eye. It is about 50 million years old. The blue light is starlight scattered by dust that the cluster probably encountered after the last vestiges of its own cloud disappeared. The Pleiades is 410 light years away.
(c) This is one of the oldest open clusters known, about 10 thousand million years old (twice the age of the Sun). It is 2700 light years away. This cluster also has only a catalogue number, M67.

2 million km

1.23 Relative sizes of stars
The stars are not all like the Sun, not in age, mass, size, or surface temperature, though they all have broadly the same initial composition, dominated by hydrogen. As a star goes through its lifetime, its size changes enormously, as shown here for a fairly average star, like the Sun. For nearly all its 10 thousand million year life, the star will look much like the Sun does now, yellowish white, and about 1.4 million kilometres in diameter. Then it will become a red giant, enormous, and with a lower surface temperature appearing orange-white. Finally it will shed a lot of mass and become a white dwarf (the tiny dot), initially with a far higher surface temperature than the Sun, but gradually cooling and reddening, until it becomes too dark to be visible to us.

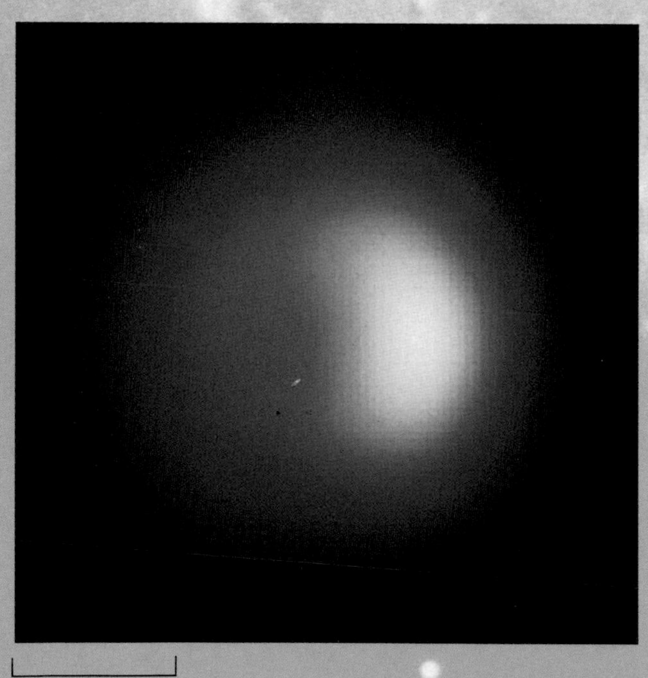

500 million km

1.24 Betelgeuse, a supergiant star
Only a few stars are both big enough and close enough for surface features to be revealed. One of these is Betelgeuse, the orange-white star in the constellation Orion (Plate 1.15), which is nearly a thousand times the Sun's diameter, and is 520 light years away. This image, obtained with red light, clearly shows a hot patch – probably a vast area of upwelling gas. Betelgeuse is about 20 times more massive than the Sun, and has thus become not a red giant, but a far larger supergiant.

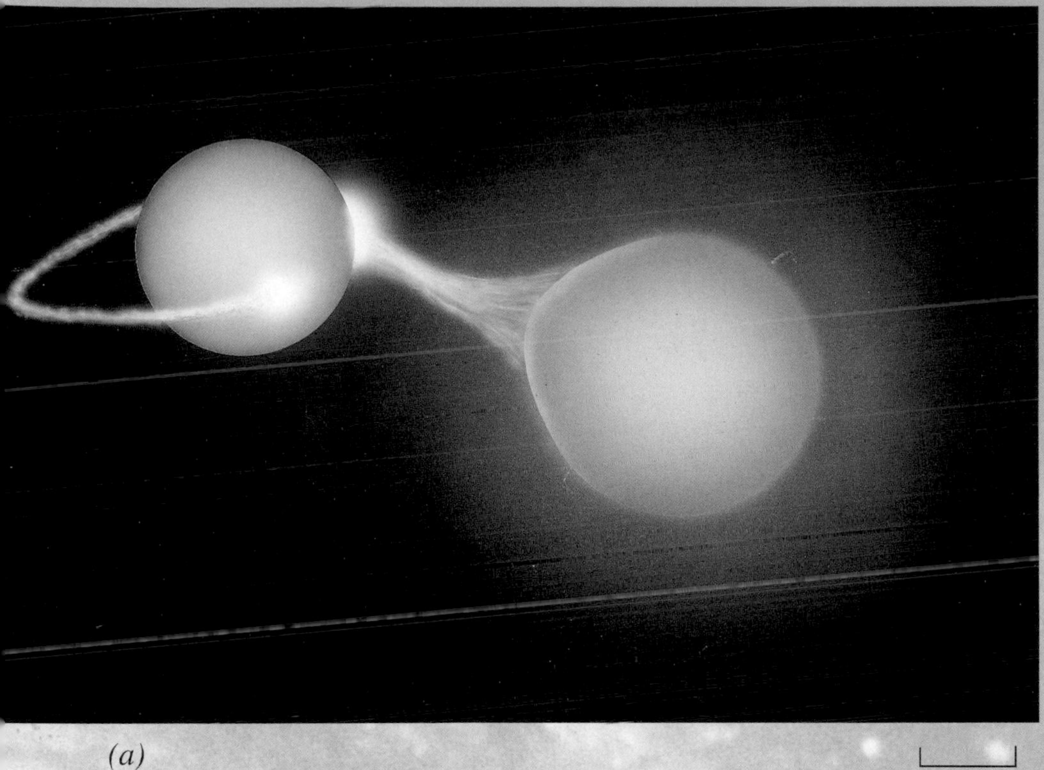

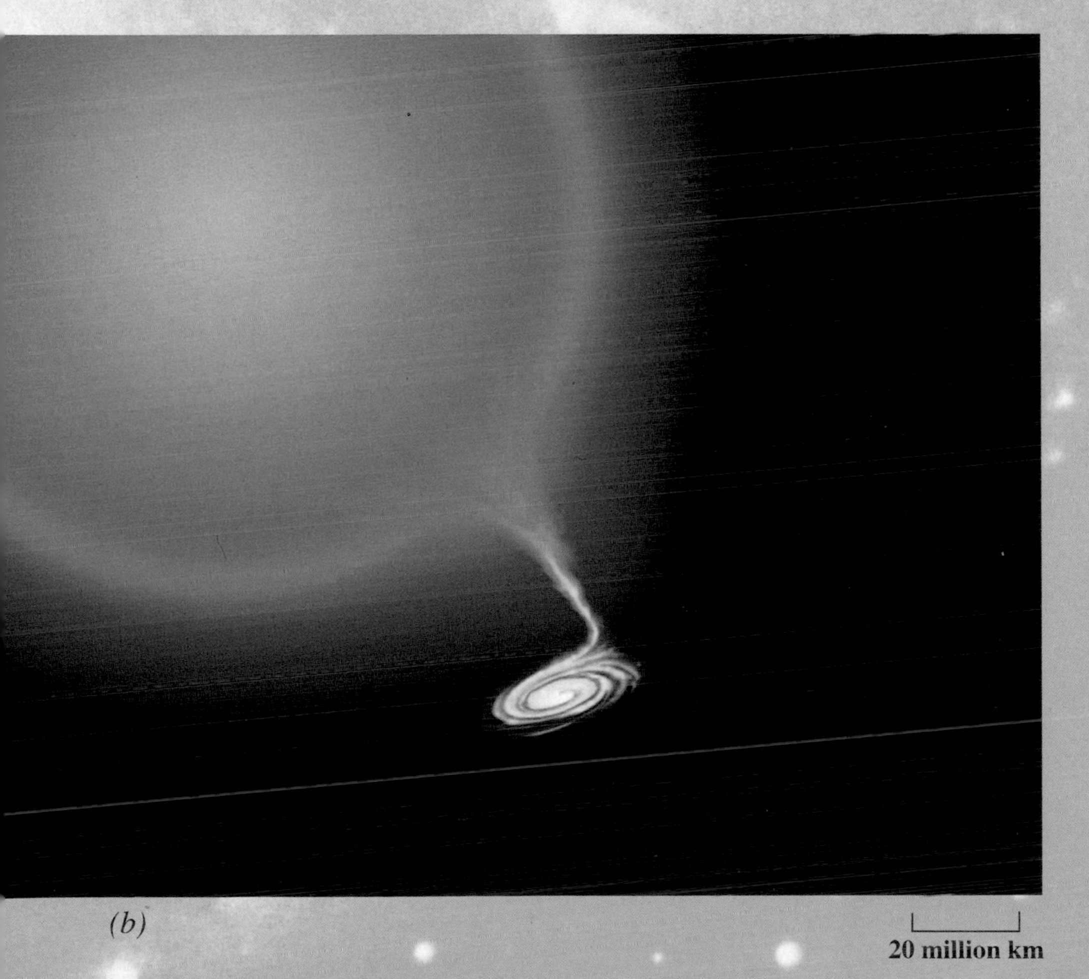

1.25 Interacting binary stars
Many stars that appear as single points of light are in fact two or more stars close together, sometimes so close that they cannot be seen separately even under the highest practical magnifications. In such cases the multiple nature of the star can be seen through effects on the stars' absorption line spectra (Plate 1.6), or, (a) as in the case of Algol, if one star passes in front of the other as they go around their orbits. When the stars are very close they can influence each other's evolution. This is the case in this artist's impression of Algol, where the more distended (yellow) star is losing mass to the other. Algol is 75 light years away. (b) In some binaries, as in this artist's impression, the smaller star can be really small – a white dwarf, or a neutron star, or a black hole. The infalling material from the larger star forms an inwardly spiralling disc, called an accretion disc; the gravitational energy released can emerge largely as X-rays, giving rise to X-ray sources.

1.26 The Toby Jug Nebula: mass loss by a red giant

During the red giant phase of a star's lifetime it will gradually lose some of its mass. As the material moves away from the star it cools, and certain elements and their compounds condense to form dust – tiny grains the size of those in talcum powder. The star thus becomes surrounded by a dust cloud, lit by the light from the red giant. This is what we see in this image of the Toby Jug Nebula, which is 316 light years away. The mass lost by red giants is one way that stars return material to the interstellar medium. The material is mainly hydrogen, which remains gaseous.

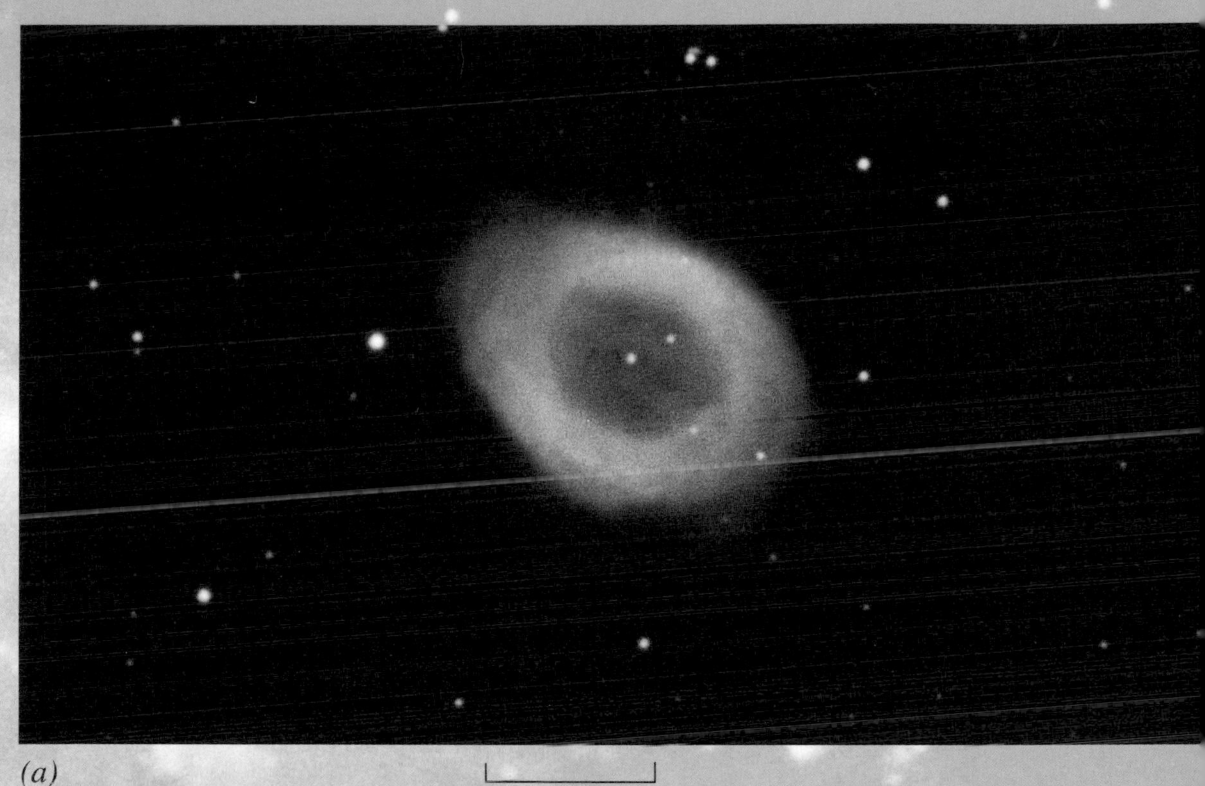
(a)

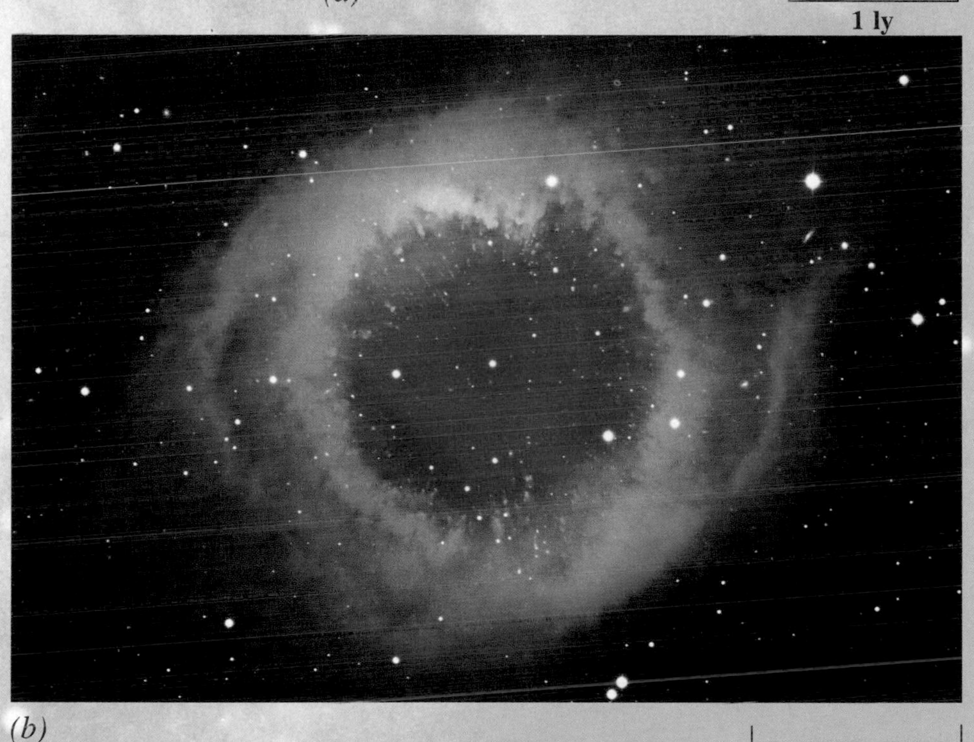
(b)

1.27 Planetary nebulas: stellar death throes
The red giant phase comes to an end when the star sheds a shell of hot material that is a significant fraction of its mass. The shell expands at several kilometres per second, cooling and thinning as it does so, until, after only about 30 thousand years, it has dispersed, replenishing the interstellar medium. (a) This is the Ring Nebula, 5000 light years away. The shell looks like a ring because we see through it most readily in the direction of our line of sight to the central star. Many planetary nebulas, under low magnification, look like the disc of a planet: hence their name. (b) This is the Helix Nebula, 400 light years away.

1.28 Supernova 1987A
Stars with masses greater than about eight times that of the Sun end their lives as supergiants that cast off nearly all of their mass in a gigantic explosion called a supernova. The majority of supernovas are too far away or too obscured to be visible to the unaided eye. Those that do become visible occur on average only once in a few hundred years. The most recent was in February 1987 – SN 1987A, 163 thousand light years away. (a) This is the sky before (right) and after the explosion. The arrowed supergiant, in about a day, was transformed into a brilliant outburst that gradually dimmed over several months. (b) Three and a half years after the supernova, its radiation illuminated this hoop-shaped ring of material that was shed by the star several thousand years *before* it exploded.

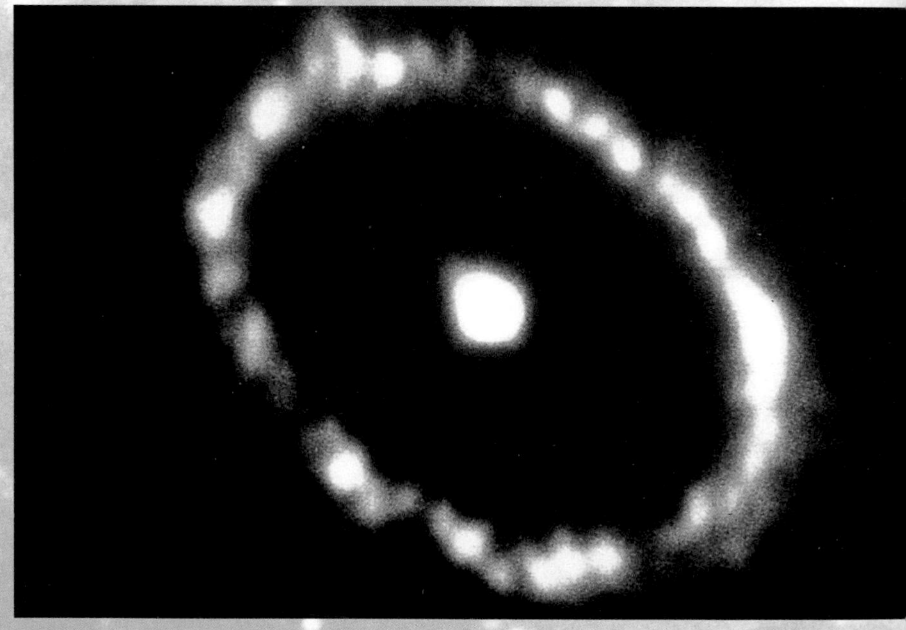

(a)

1 ly

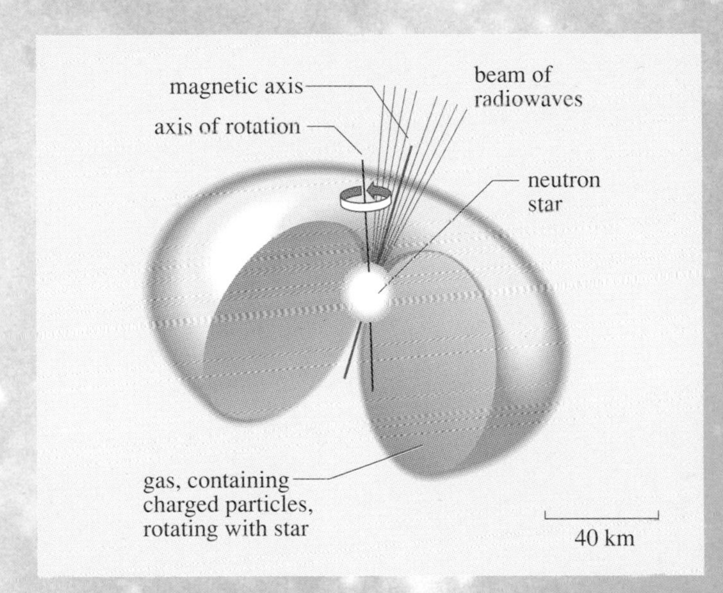

(b)

1.29 The Crab Nebula: a supernova remnant
Many years after a supernova, we are able to observe the tortured material thrown out at thousands of kilometres per second. This material is the supernova remnant. (a) The Crab Nebula is one such remnant, 4000 light years away. The supernova that gave rise to the Crab Nebula was observed by the Chinese in AD 1054. The star just below the centre of the image is all that remains of the exploded star, though it still energizes the remnant. The star is a pulsar, an extremely dense, rapidly rotating body, only about 10 kilometres in radius! Pulsars get their name from the regularly spaced pulses of radio waves that they emit, typically several per second. They consist largely of atomic particles called neutrons, so they are also neutron stars. (b) This is an artist's impression of a pulsar, about the size of the Crab pulsar. The pulses occur because a beam of radio waves sweeps across our telescopes once per revolution of the star. In reality, the torus of gas would extend farther than shown here.

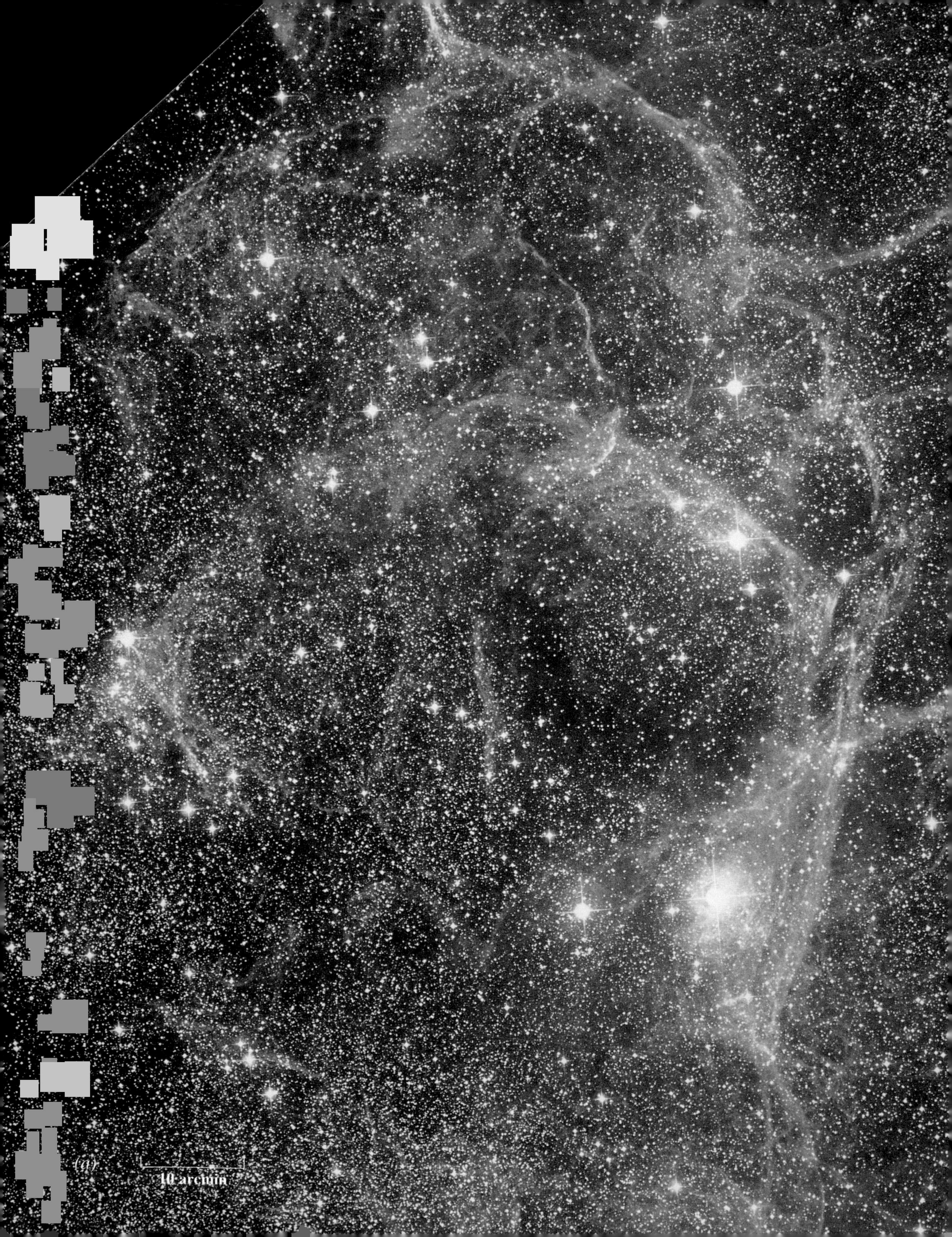
(a)
10 arcmin

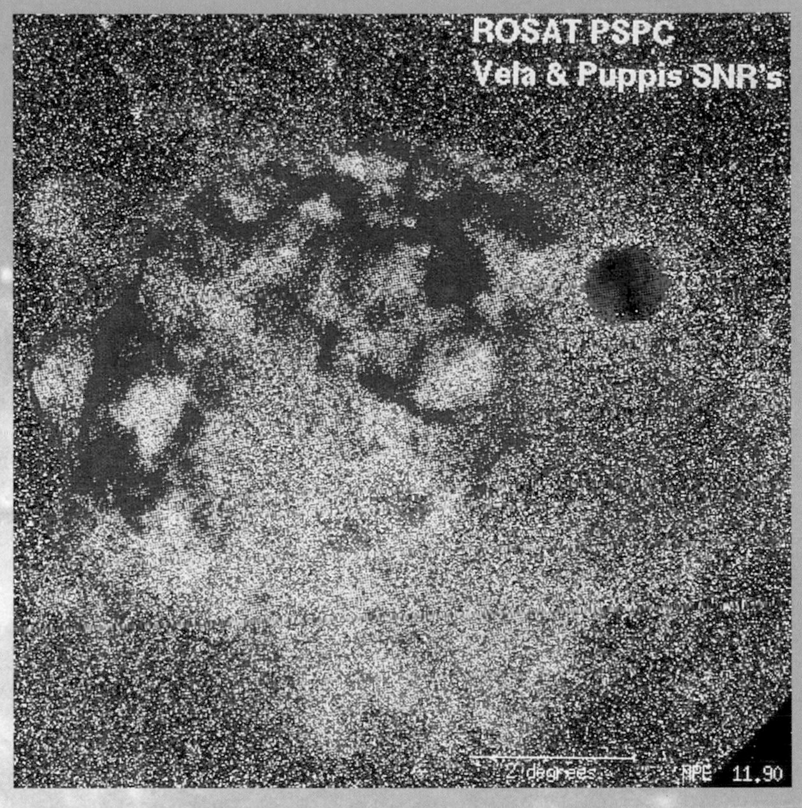

(b)

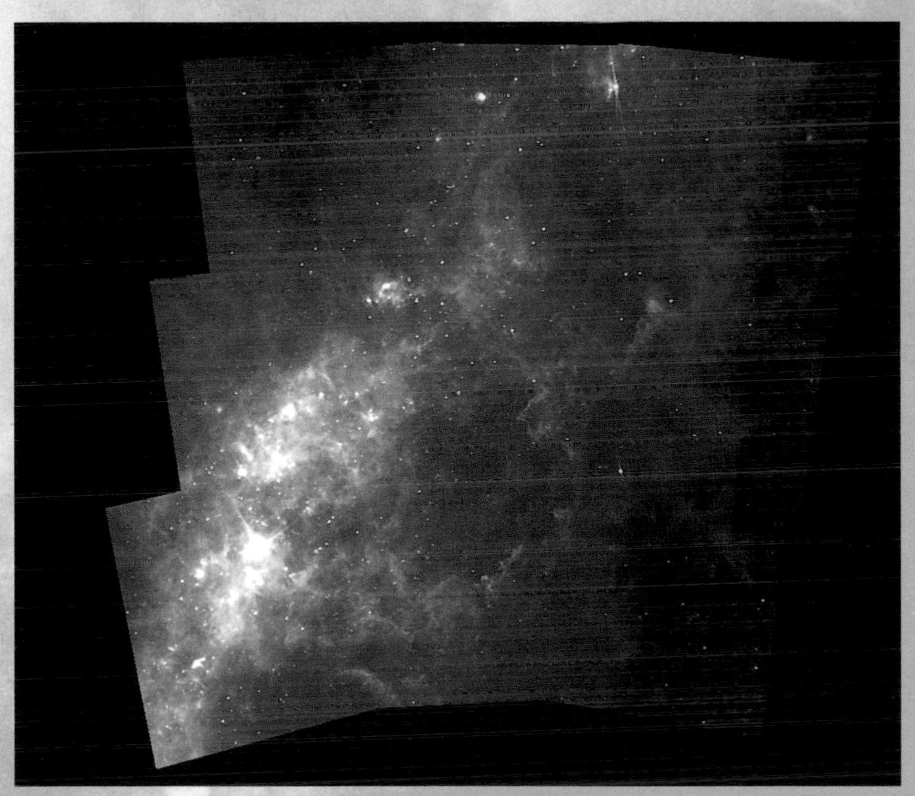

(c)

5 degrees

1.30 The Vela supernova remnant
The Vela supernova remnant is one of the nearest. It is about 11 thousand years old and so is large – about 2300 light years across. It ranges from about 300 light years to about 2600 light years from us. Approximately in its centre there is a pulsar. (a) This image at visible wavelengths is of only a small portion of the Vela remnant; it spreads over 10 times further across the sky, though it is so faint that it can't be seen with the unaided eye. The light is produced when the material in the remnant ploughs into pre-existing interstellar matter. This collision produces radiation over a wide range of wavelengths, from (b) X-rays, to (c) radio waves. A much greater fraction of the whole remnant is shown in these non-visible images.

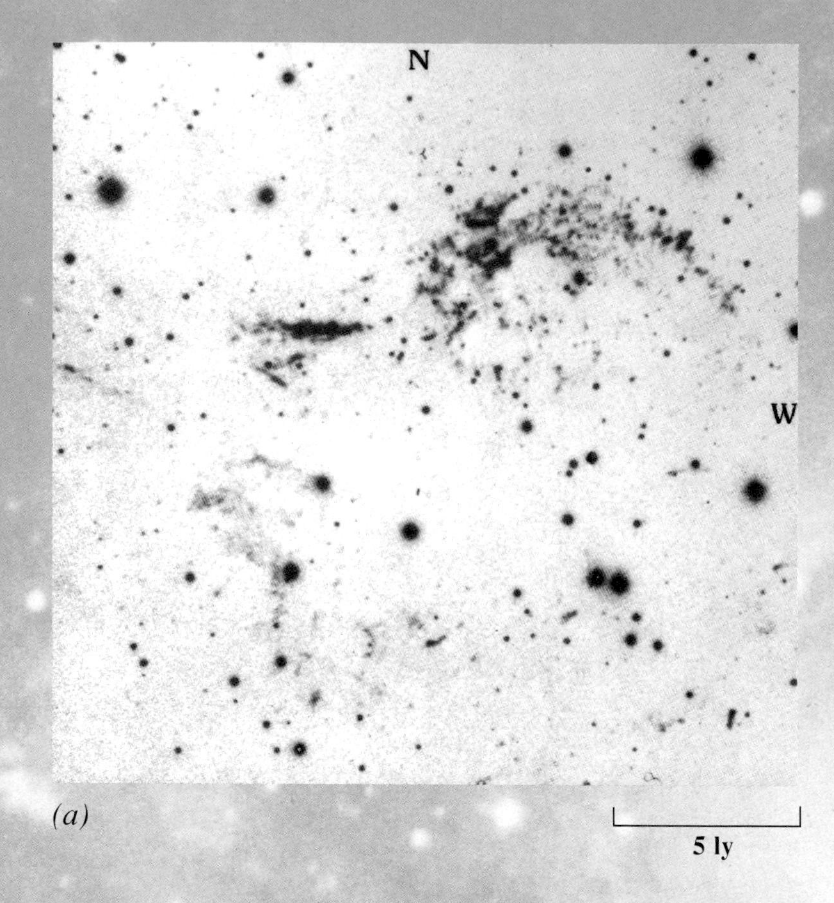

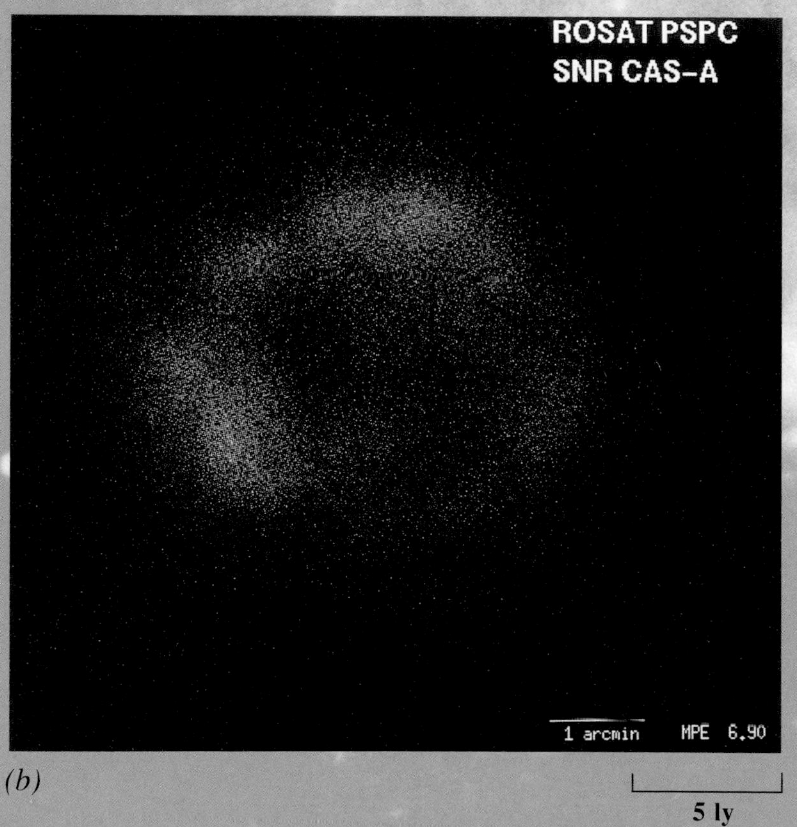

1.31 The Cassiopeia A supernova remnant
The supernova that gave rise to the Cassiopeia A supernova remnant occurred about 300 years ago, 11 thousand light years away. No pulsar has been detected. (a) Many remnants are not easy to pick out at visible wavelengths, as is apparent from this negative image of Cassiopeia A. (b) With X-rays, Cassiopeia A, like some other remnants, is easier to pick out. (c) At radio wavelengths, it is even easier. Indeed, most of the several hundred known remnants have been discovered by radio telescopes.

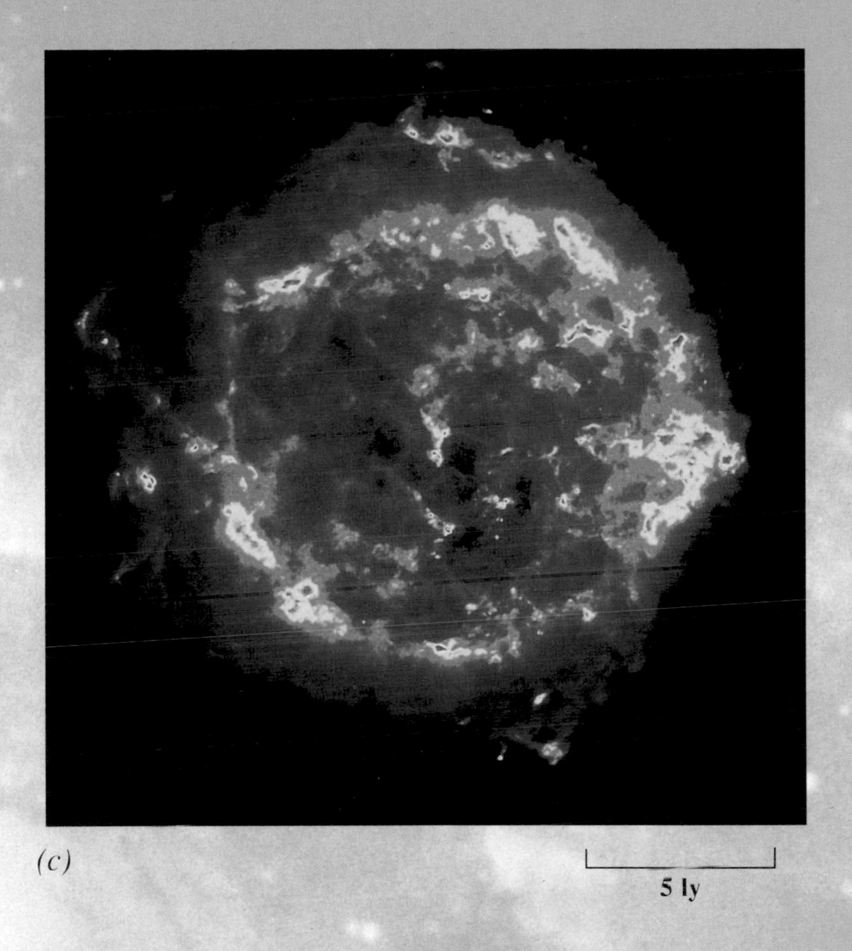

(c)

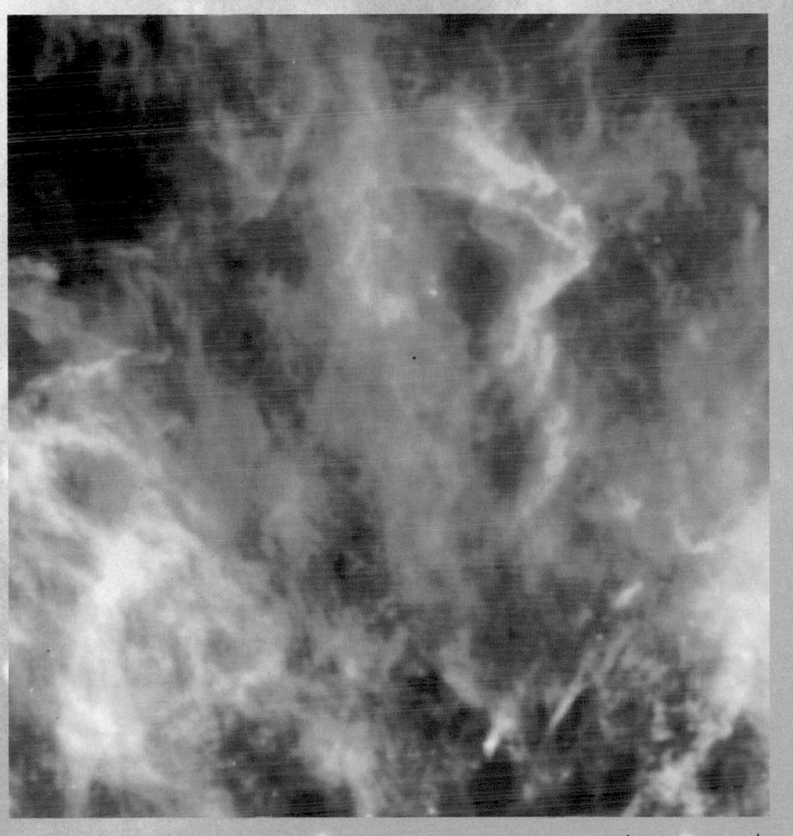

1.32 Interstellar cirrus
What happens to all the material that stars return to the interstellar medium? Ultimately, some of it forms new dense clouds, leading to the birth of a new generation of stars. An intermediate stage might be cool, very thin wispy clouds, called interstellar cirrus. Though widespread, they were not discovered until 1984, in images made by an infrared telescope in space, called IRAS, which detected the infrared radiation from the cold dust in the clouds. The infrared image here shows cirrus a few hundred light years away.

1.33 Some important telescopes
(a) This is the New Technology Telescope of the European Southern Observatory, completed in 1989, and located at La Silla in Chile, 2350 metres above sea-level where atmospheric degradation of image quality is comparatively slight. The telescope operates at visible wavelengths, and some way into the infrared. This telescope was the first to have *active optics*, an automatic system of keeping the optics in optimum shape, thus yielding images far crisper than any obtained earlier at visible/infrared wavelengths. (b) At infrared wavelengths, except those nearest to the visible, there is an advantage in getting above our atmosphere, which is partially opaque and also glows at such wavelengths. This infrared telescope is IRAS – the InfraRed Astronomy Satellite – developed jointly by the UK, the USA and the Netherlands, and launched into Earth orbit in January 1983. It operated for nearly a year, and made many important discoveries (e.g. Plate 1.32).

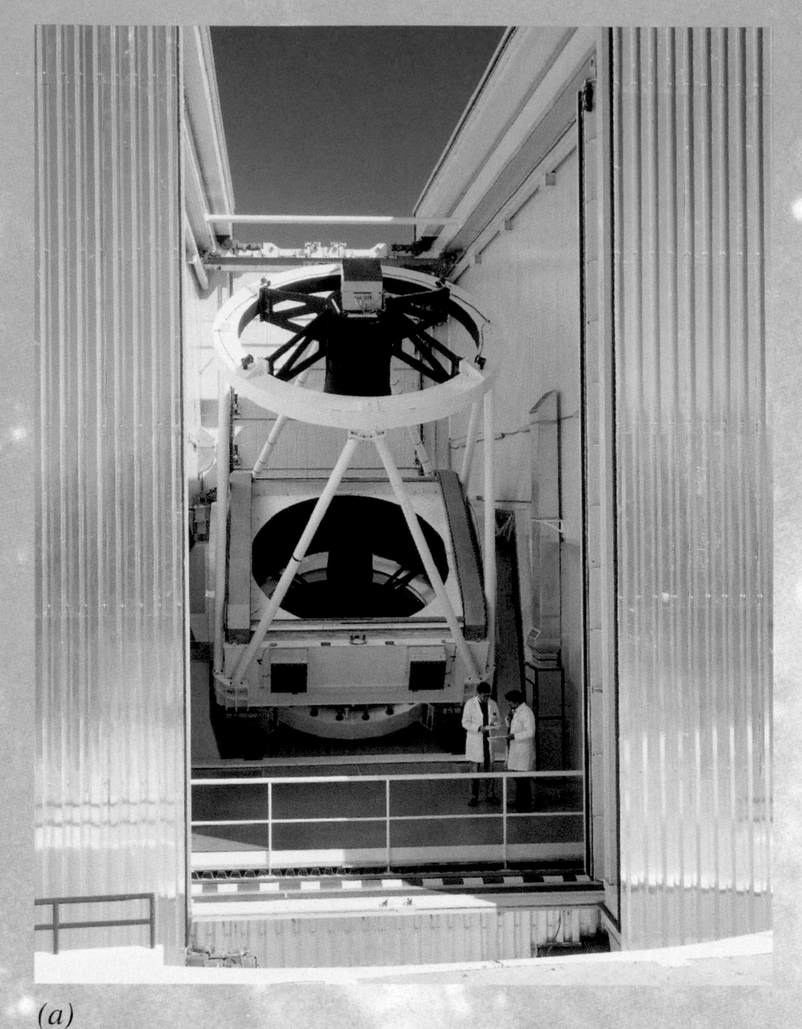

(a)

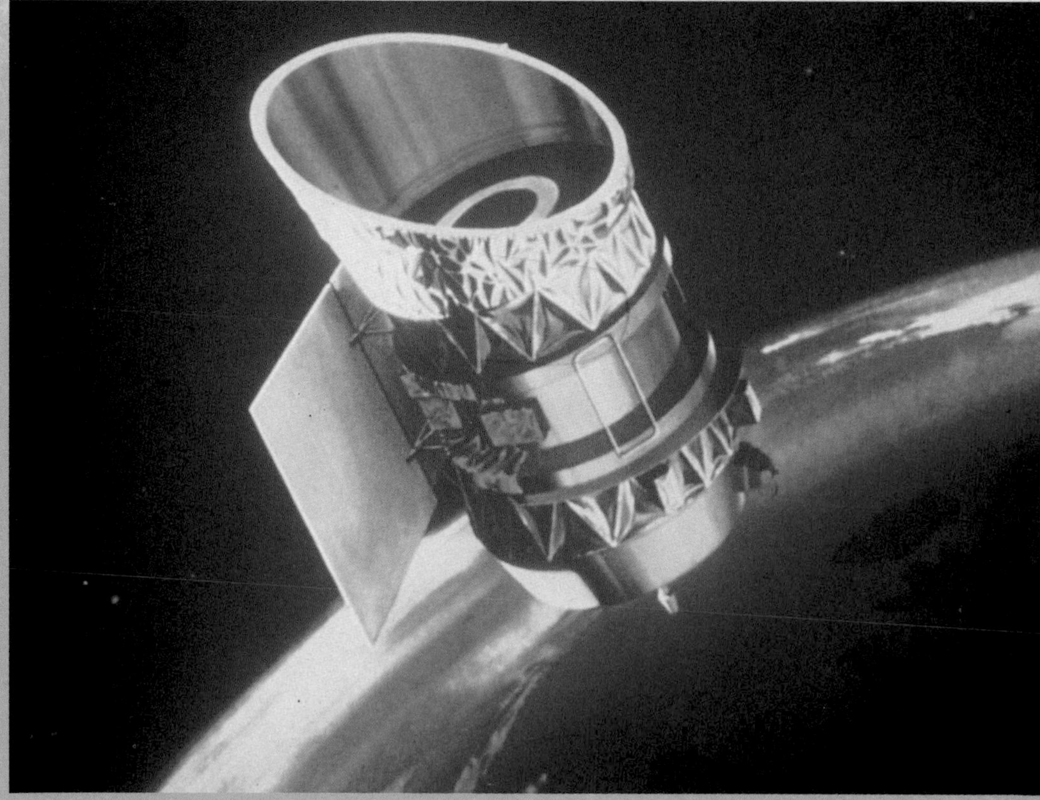

(b)

1 metre

(c)

1 metre

(c) X-ray telescopes have to be placed above our X-ray opaque atmosphere. This is ROSAT, a Germany–USA–UK X-ray satellite, launched into Earth orbit in June 1990.
(d) Radio telescopes can operate at ground level. Often they consist of an array of collecting dishes, rather than a single dish. By combining the output from all the dishes, much finer detail can be obtained than with each dish alone: the more spread out the array, the finer the detail. This is the Very Large Array, located in New Mexico, and completed by the USA in 1981. Each dish is 25 metres in diameter. There are 27 dishes, and they can be moved apart to cover an area 32 kilometres across.

(d)

Part 2

Our planetary system

We now retreat from interstellar distances to take you on a tour of our nearest neighbours (and indeed the Earth itself). The Sun is much the most massive body in the Solar System, constituting about 99.8% of its total mass. You have already seen several images of this, so here we shall examine the planets and smaller bodies that make up the rest of the Solar System. Although these bodies are small in comparison with other observable objects in the cosmos, they are particularly fascinating because they offer insights into the variety of possible worlds other than our own.

Our planetary system consists of nine major planets, a large number of natural satellites in orbit about these planets, nearly 5000 known minor planets or asteroids, and a large, indefinite number of comets. Some information about these bodies and the spacecraft that have explored them is listed in the tables at the end of this book. The diagram opposite shows the layout of the Solar System in oblique view. All the planets orbit the Sun in the same direction: anticlockwise as seen from above the Sun's (or the Earth's) north pole. These orbits are ellipses, lying roughly in the same plane, though except for Mercury and Pluto their shapes are very nearly circular. Most of the planets spin on their axes with the same anticlockwise sense of rotation. The exceptions are Venus, which spins very slowly backwards, and Uranus and Pluto, which are tipped on their sides. The orbits of all planetary satellites lie close to the plane of their planet's equator, and most satellites travel in the same direction as their planet's spin. The asteroids are concentrated in the region between Mars and Jupiter and orbit the Sun in the same direction as the planets, although they are confined less tightly to the main plane of the Solar System. Comets can have extremely elongated orbits, inclined at any angle to the plane of the Solar System. The shared sense of rotation of almost everything except comets is among the evidence used to suggest that the Solar System condensed (just over 4500 million years ago) from a rotating disc-like cloud of gas and dust known as the solar nebula (Plate 2.1).

The diagram also shows that four of the outer planets (Jupiter, Saturn, Uranus and Neptune) are giants, considerably bigger than the rest. This does not mean simply that the solar nebula was fatter in this region, rather that its temperature was lower here. In the inner reaches the only substances to condense in large amounts were metals and the silicate minerals that make up rock, and this is responsible for the comparatively high densities of the four inner planets, and the Earth's Moon too, which are collectively known as the terrestrial planets. At Jupiter and beyond it was cold enough for water and other volatile substances (such as ammonia and methane) to condense, and because of this the giant planets grew so big that they were able to capture substantial amounts of gaseous hydrogen and helium directly from the nebula. Some of the icy and rocky material that gathered in the vicinities of each of these giant planets appears to have avoided capture onto the planets themselves and went instead to form their satellites. In the sparse outer reaches of the solar nebula, beyond the orbit of Neptune, relatively little material was available for the planet-forming process, hence the small size of the outermost planet, Pluto.

When we look at the giant planets, we see atmospheres driven by the planetary rotation (which is considerably faster than the Earth's rotation) and stirred up by heat from the Sun. We can see atmospheric processes on many smaller bodies too, but these also have solid surfaces that you could walk on, if you could get there.

Those planetary bodies that are inert are peppered with craters as a result of impacts by Solar System debris over the ages, but many worlds have long histories of activity powered by the escape of heat from their interiors, and so parts of their surfaces are buckled, and flooded by volcanic outpourings.

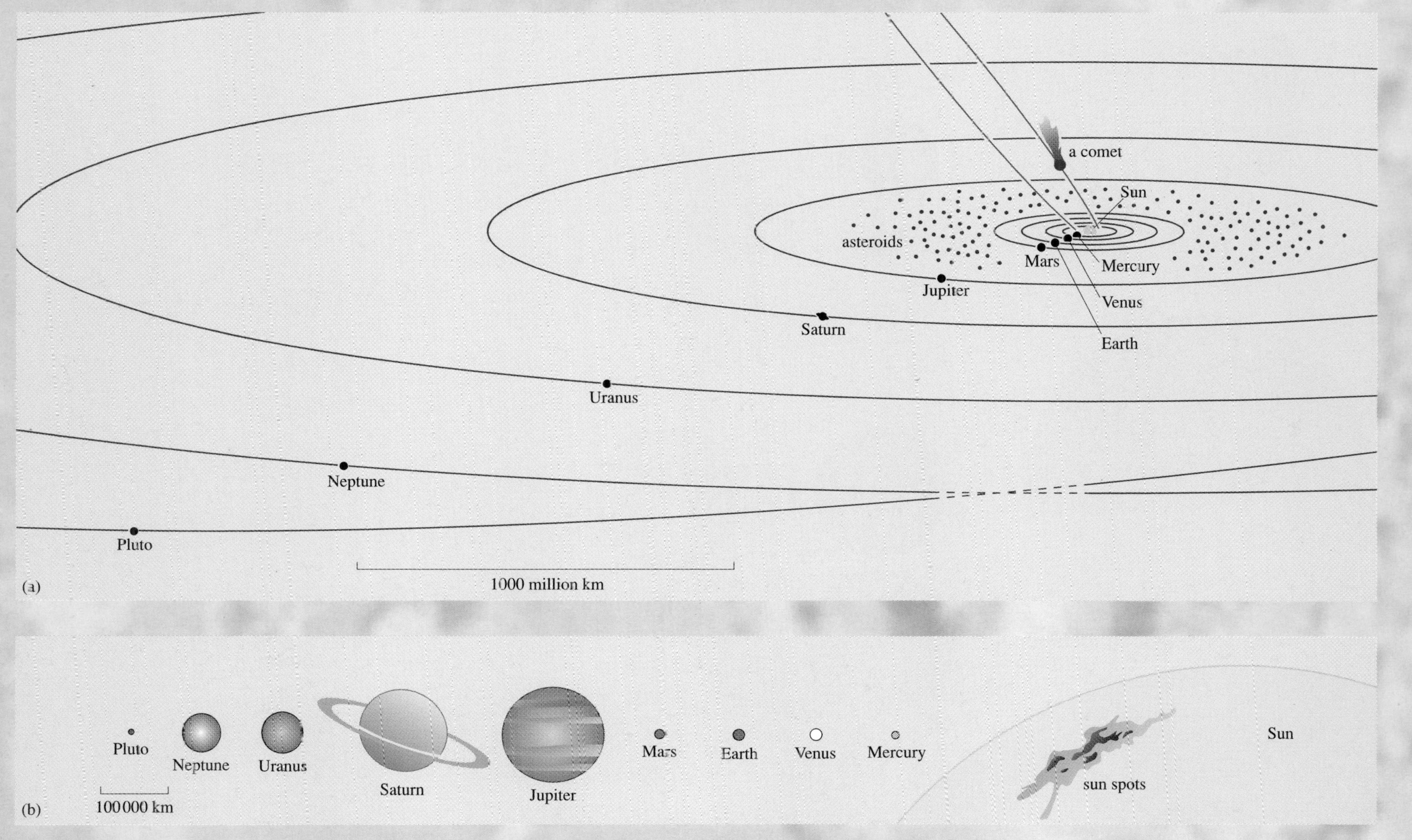

Top: the orbits of the nine major planets, as seen looking obliquely southwards from outside the Solar System. Such a view distorts the shapes of the orbits, which, apart from those of Mercury and Pluto, are virtually circular. The asteroids and the orbit of a single comet are shown schematically. Bottom: the Sun and the nine major planets, showing their true relative sizes

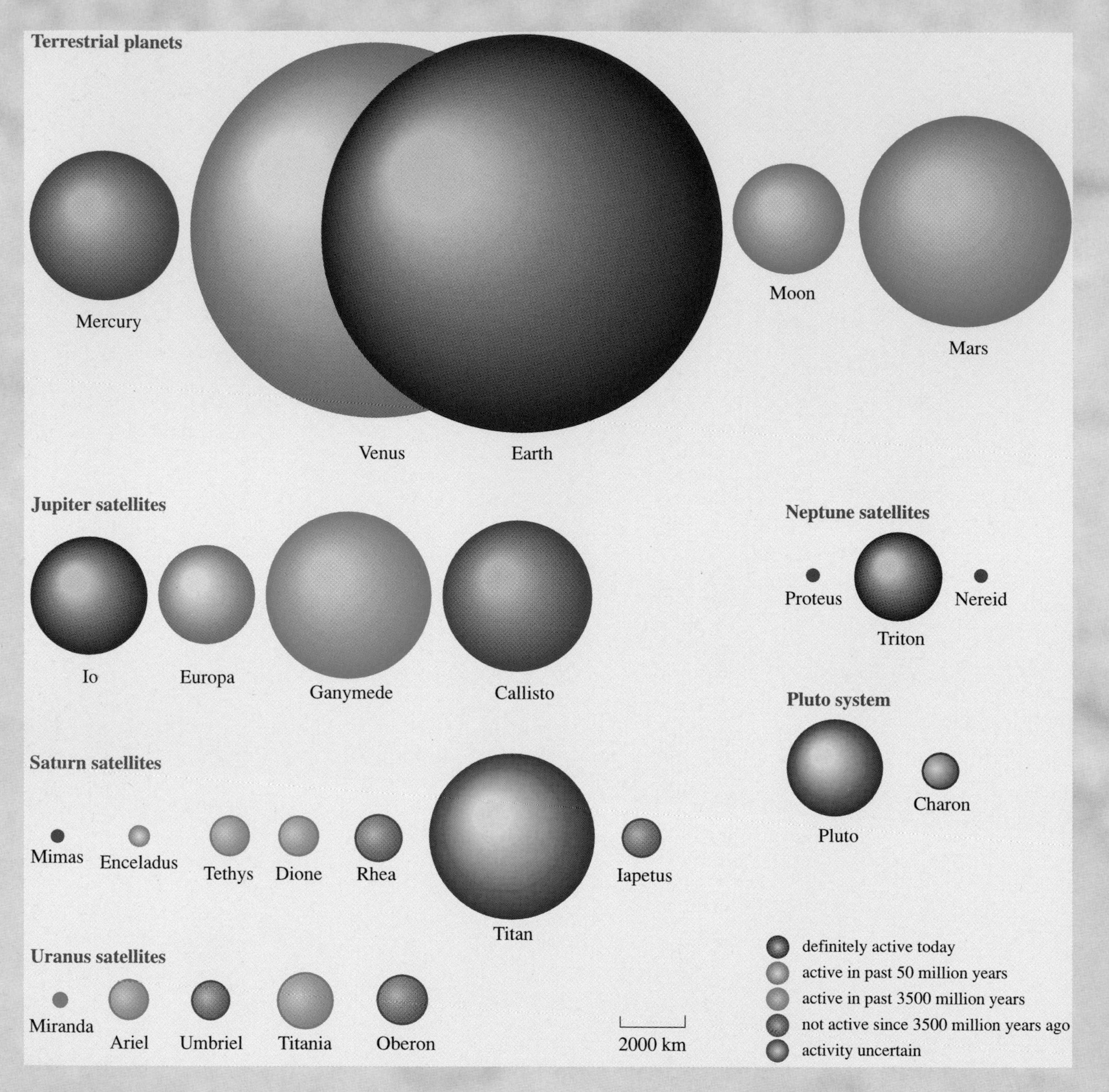

The terrestrial planets and all the planetary satellites more than about 200 kilometres in radius drawn to scale and colour-coded according to how recently their surfaces have been affected by volcanism, fracturing or buckling.

Variations in the effectiveness of the processes of heat generation within planetary bodies are indicated in the diagram above. This shows all the main rocky and icy bodies in the Solar System drawn to the same scale, but colour-coded according to how recently they have been deformed, or resurfaced by volcanism. Of the three bodies definitely active today, the Earth is a rocky body with the greater part of its heat coming from the decay of radioactive elements, Io is a rocky body that is kept active by tidal heating, and Triton is an icy body with geysers that may be powered by sunlight, trapped greenhouse-fashion by a blanket of nitrogen-ice. How recently a terrestrial planet has been active depends largely on the amount of heat-producing radioactive elements that it contains (which in turn depends on its size), whereas activity among most of the satellites of the outer planets appears to depend on the history of their tidal interactions with neighbouring satellites and the planet that they orbit.

A note on planetary interiors

Descriptions of the main bodies with solid surfaces in the Solar System often divide them into two separate groups, rocky ones and icy ones, according to the nature of the material that makes up their outer layers. The rocky bodies are the terrestrial planets and Io (Jupiter's innermost large satellite), whereas the icy bodies are all the other large satellites of the outer planets and the planet Pluto. However, to categorize planetary bodies in this way is to overlook important similarities in the nature of the processes that occur upon them and within them. The diagram here shows the interiors of two planetary bodies, one of each class. The rocky body (which could be the Earth) has a dense core mainly of iron; almost everything else is made of silicates. The outermost layer (about 100 kilometres thick for the Earth) behaves rigidly and is called the lithosphere (meaning 'rocky shell'). Between this layer and the core, thanks largely to internal heat generation, the silicate material is so hot that, even though it is not quite molten, it can flow at a rate of a few centimetres per year. This weak interior is stirred up by convection currents that transport heat towards the surface. Where upcurrents (plumes) hit the base of the lithosphere they may cause volcanism at the surface, and where currents diverge or converge the lithosphere may be fractured or buckled. An icy body functions in precisely the same way. Any core is likely to be of rocky material, but this is deeply overlain by ice, consisting of frozen water mixed with a proportion of volatile substances, such as ammonia and methane, that increases with distance from the Sun. The outermost layer of ice is so cold that it is very rigid and behaves exactly like rock on Earth, forming an icy lithosphere. Below this, provided that there is a supply of internal heat, the ice becomes mobile enough to convect, leading to the same range of styles of surface deformation and volcanism as displayed by the terrestrial planets, except that the volcanism involves melts derived from ice rather than molten rock.

What we see when we look at a solid planetary body depends on whether or not there is enough heat escaping from the interior that the lithosphere remains thin enough for it to be punctured by volcanism and deformed by the underlying convection currents. Bodies whose lithospheres have grown too thick for these processes to be effective show progressively less evidence of volcanism and deformation; these surface traces gradually become obliterated by craters induced by the occasional impact of asteroid and comet debris that continues even today.

Cut-away views to show the interiors of a rocky planet (such as the Earth) and an icy body (like Pluto and most of the satellites of the outer planets). A convecting layer, of hot rock in one case and 'warm' ice in the other, transports heat outwards. If the outer rigid shell, or lithosphere, of either type of body is thin enough it will shows signs of deformation and volcanism.

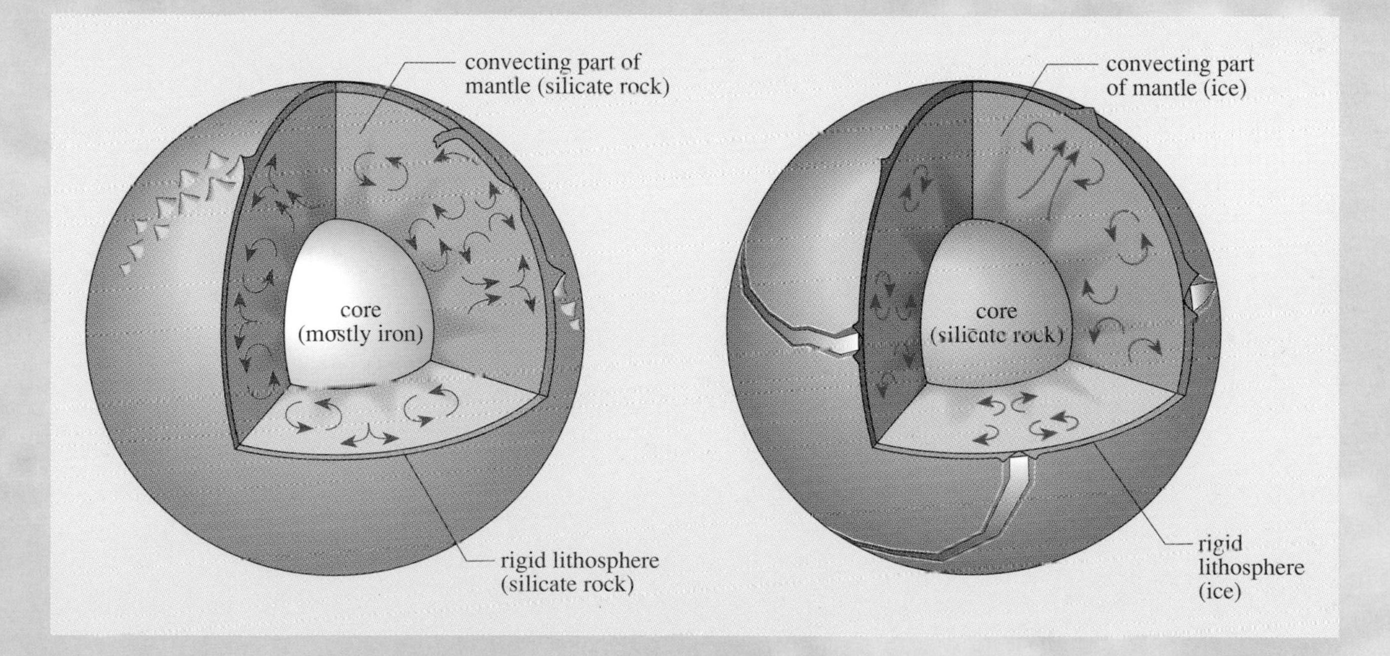

The giant planets are a different matter, as shown in the last diagram. Their hydrogen-rich atmospheres are very deep, extending inwards for considerable proportions of the planetary radius, and the pressure and temperature become immense in the interior. Within Jupiter and Saturn the pressure squeezes hydrogen atoms so close together that they behave like liquid metal. Any solid surface is buried so deeply that we are unlikely ever to discover what it looks like.

The plates in Part 2 are arranged in an order chosen to illustrate the progression from dead, inactive worlds through to those that are active today, so as you turn the pages you will find rocky planets and icy satellites mixed together. Having reached the most active world in the Solar System from the point of view of internally driven heat processes (Io), we then look at processes occurring in planetary atmospheres, first the atmospheres of planets like our own, and then the four giant planets where all we are ever likely to see is the outermost part of their immensely deep atmospheres. Your tour of the Solar System ends with a look at some of its smaller members: asteroids, a comet, several meteorites, and (perhaps the most beautiful sight of all) Saturn's rings, which are composed of a myriad of icy fragments girdling the planet.

The Solar System is small in comparison with interstellar distances. The orbits of the outermost planets are barely a thousandth of a light year across, and even the most far-travelled comets probably originate no more than a light year away. However, be it ever so humble, it is our home, and we can be justifiably proud that we live in such a fascinating neighbourhood.

Cut-away views to show the interiors of Saturn (left) and Uranus (right) (not to same scale).

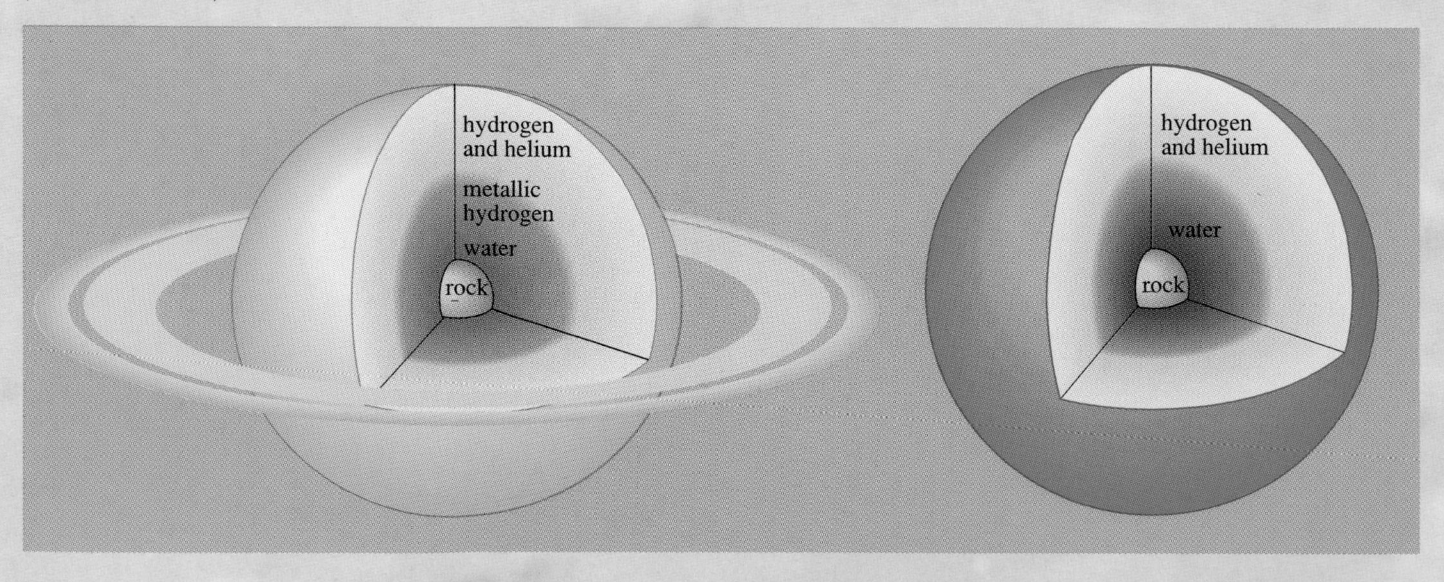

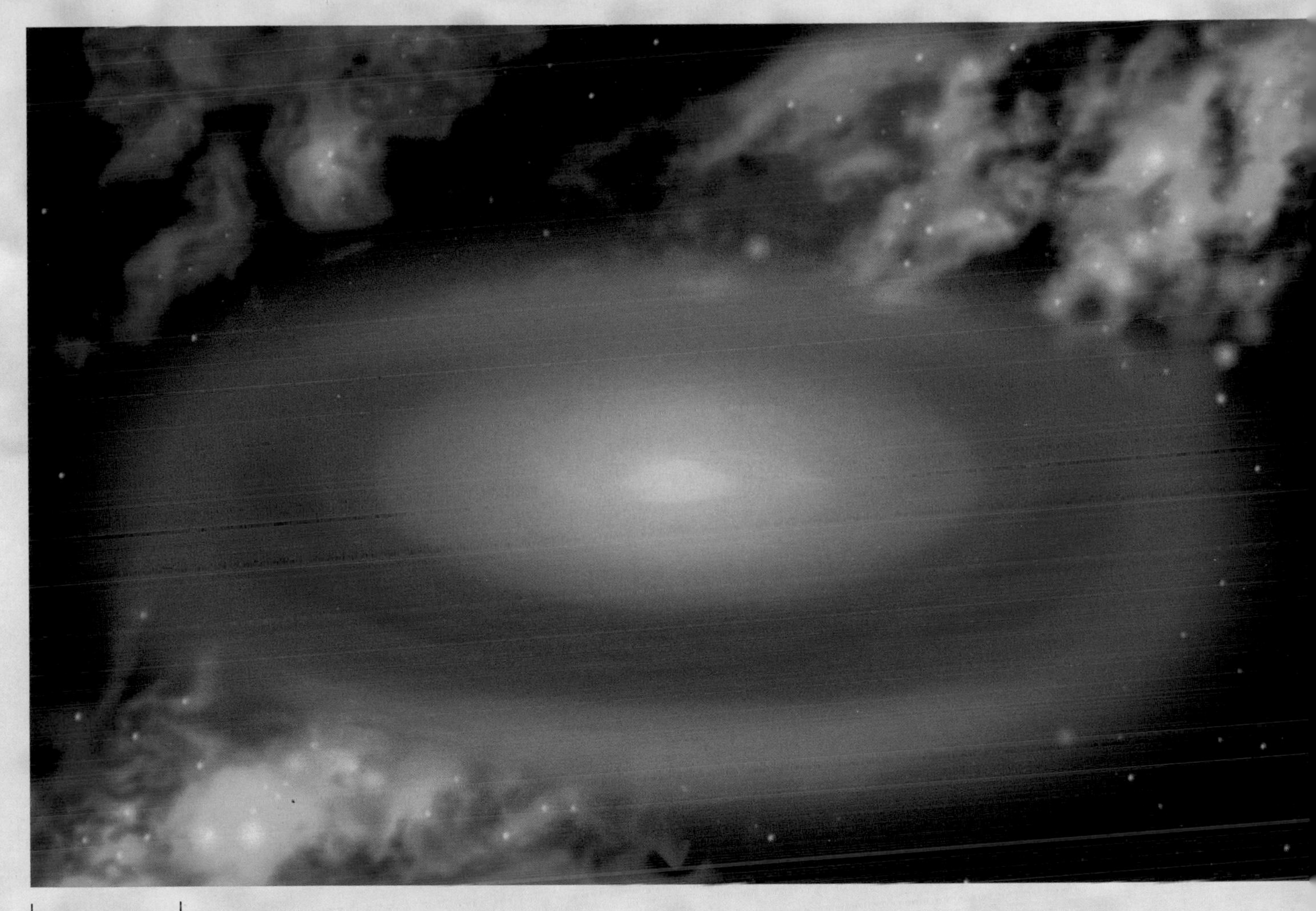

1 billion km

2.1 As it was in the beginning?
The solar nebula
An artist's impression of the Solar System at its birth. Although many details are still in dispute, most scientists agree that the Solar System formed within a rotating disc of gas and dust (known as the solar nebula) that developed from a contracting part of an interstellar cloud nearly 4600 million years ago. At the stage illustrated, the central, densest part of the nebula, which will shortly become the Sun, is shining only because it has grown hot by release of gravitational energy as it has contracted. Metals and rocky material are condensing as dust grains within the hottest part of the nebula (closest to the Sun) and these are joined by ices and other more volatile substances in the cooler, outer regions. These grains stick together when they come into contact, and within about a million years most of this material will have been swept up to form the planets.

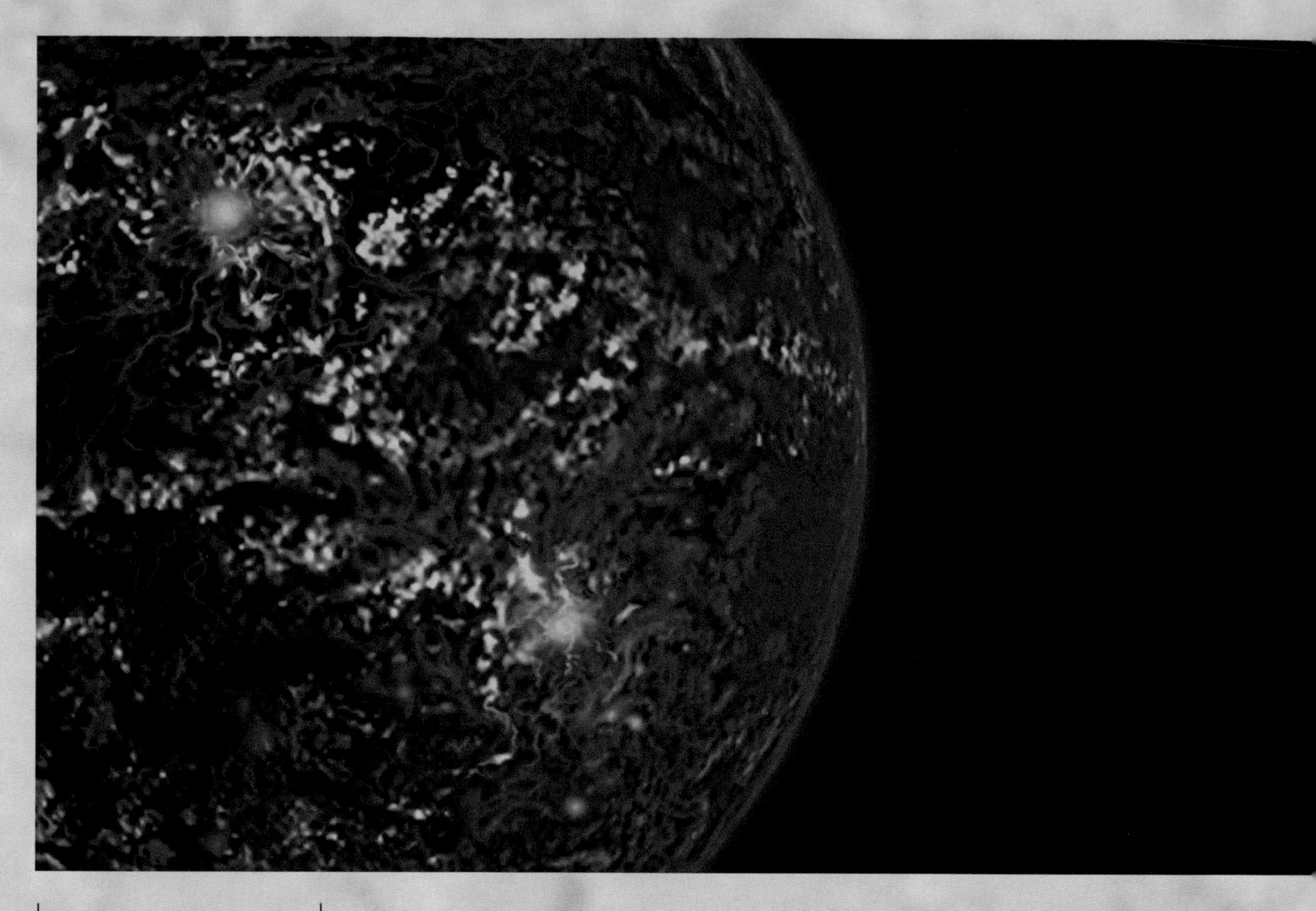

2.2 A young planet
This artist's impression shows the Earth as it may have appeared about 4550 million years ago. The outer part of the whole globe has been melted by the energy released during a recent collision with a body of maybe one-tenth the total mass of the planet (known as a planetary embryo). Each terrestrial planet is thought to have grown as a result of several such collisions between planetary embryos. The Moon may be a by-product of a glancing blow from the last planetary embryo to strike the proto-Earth. In this view, the global 'magma ocean' of molten rock is covered by a chilled crust, torn apart from below in several places as a result of vigorous convection and de-gassing, and punctured by the impact of asteroid-sized chunks of meteoritic debris.

100 km

2.3 Two views of Mimas
Mimas is an icy satellite of Saturn, only slightly smaller than Enceladus (Plate 2.21), and with a similar density. However, the histories of these two bodies have been very different. As these two views reveal, Mimas has an ancient surface dominated by impact craters and lacking traces of recent geological activity. The largest crater, named Herschel after Sir William Herschel who discovered Mimas in 1789, is 130 kilometres across. This is evidently the result of a collision with a body some 10 kilometres in diameter and travelling at several tens of kilometres per second. If this had been slightly larger or travelling a little faster, the collision would have broken Mimas into several fragments.The debris would probably have been able to re-assemble, thus forming the satellite anew. It is likely that such a fate may have befallen several of the innermost satellites of Saturn and Uranus. These pictures, like most of the other plates of the outer planets and their satellites were transmitted to Earth in digital form by one of the two Voyager spaceprobes (see inset) that explored the outer Solar System between 1979 and 1989.

500 km

2.4 Mercury
A view of part of the Sun's closest planet, imaged by the remarkable spaceprobe Mariner-10, which made three close passes of Mercury in the year beginning March 1974. This revealed Mercury to be a densely cratered world, at first sight not dissimilar to the Earth's Moon. The left-hand part of this picture includes the sunlit half of a major impact structure, the Caloris Basin, dating back some 4000 million years. The interior of the basin appears to have been flooded by lava, and then fractured during subsidence. Beyond the outer ring of the basin are many secondary craters produced when ejecta from the Caloris Basin struck the surface. Although considerably smaller than the Earth, Mercury is almost as dense and so, allowing for internal compression, it seems to contain a greater proportion of iron. This is most simply explained if high temperatures in the inner part of the solar nebula stopped the less refractory elements from condensing; alternatively the original rocky outer layer may have been stripped away in a collision between two planetary embryos while Mercury was being formed.

20 km (background)

2.5 The lunar crater Copernicus
An oblique view, photographed by an astronaut on Apollo 17. This shows several of the main features characteristic of impact craters that distinguish them clearly from volcanic craters: the crater floor is lower than the ground-level beyond the crater (in this case by about 3 kilometres), the crater is surrounded by a blanket of ejected material, and the inner wall of the crater has slumped into a series of terraces (evidently soon after the impact-driven excavation was complete). The projectile that formed Copernicus, some 900 million years ago, was probably a small asteroid or comet nucleus about 5 kilometres in diameter that struck the lunar surface at a speed of 10 to 30 kilometres per second. It probably took less than 2 minutes for the crater to be excavated.

2.6 Aristarchus crater and lunar rilles
An Apollo 15 astronaut's view looking south-eastwards across the Mare Imbrium region of the Moon. The bright, prominent impact crater towards the upper left is Aristarchus, one of the youngest craters on the Moon (probably around 400 million years old). This is one of the most common sites of 'transient lunar phenomena', temporary changes in brightness or colour, reported by telescopic observers. There is no generally accepted explanation for these, but they are unlikely to be emissions of volcanic gas as was once thought. Closer than Aristarchus you can see traces of a much older crater, named Prinz, that was flooded by the lavas that spread across the mare basins some 3500 million years ago. The sinuous channels are rilles cut by flowing lava towards the end of this event.

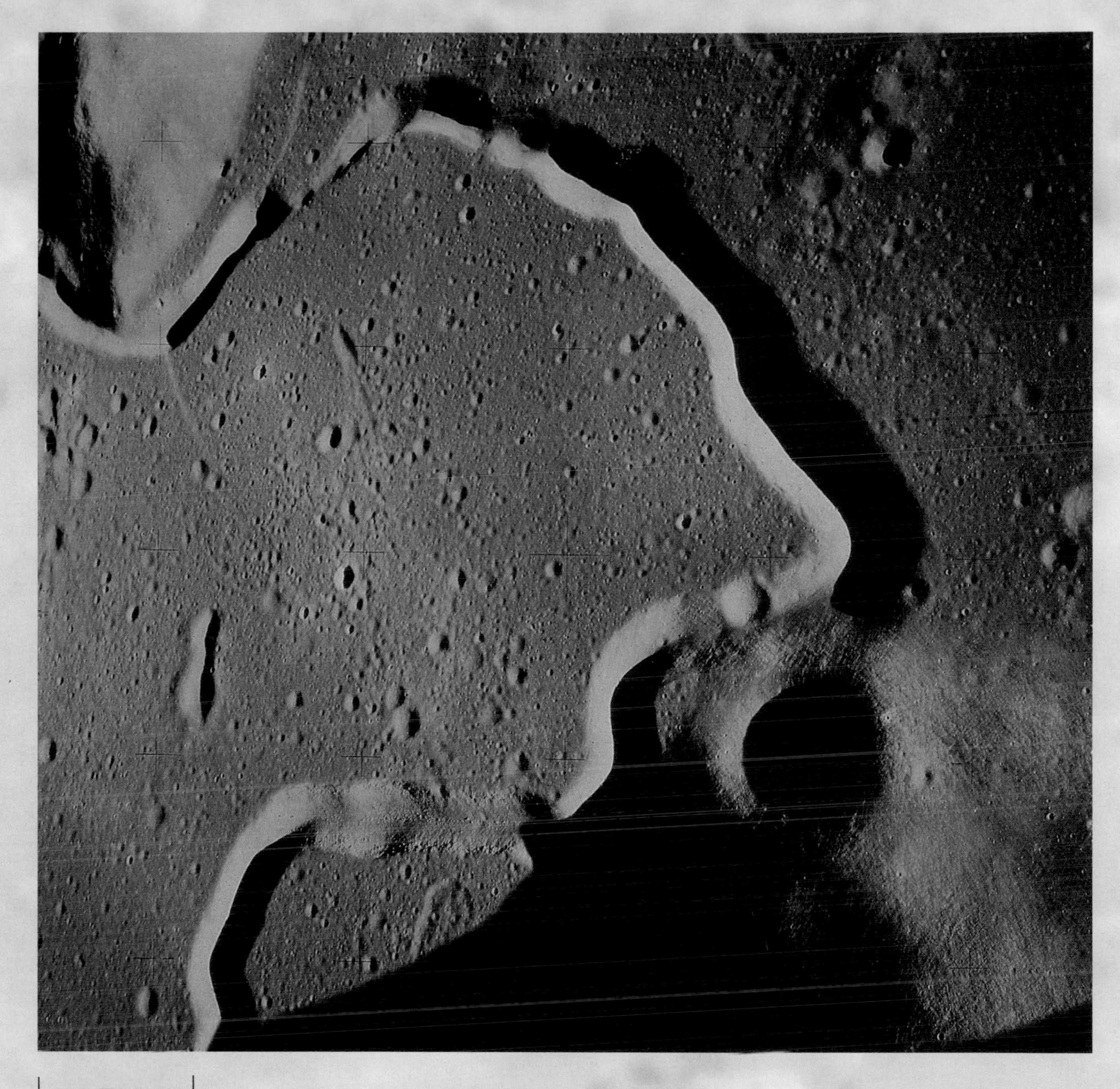

2.7 Hadley Rille
The site of the fourth manned Moon landing, as it appeared during the descent of the Apollo 15 lunar module. Hadley Rille is one of many sinuous channels within the lavas that occupy the lunar maria, and is generally interpreted as a channel formed by flowing lava. Rather than being valleys eroded into a lava surface, most rilles are probably channels along which lava continued to flow while becoming solidified to either side. Many may once have been roofed over as 'lava tubes', whose roofs collapsed after the lava drained out.

50 km (foreground)

2.8 Miranda, with Uranus in the background
A view across the irregular and faulted terrain of the small icy world of Miranda. Uranus is in the background surrounded by a ring of debris, which may be composed of the remains of one or more small icy satellites of Uranus that were shattered by collisions.

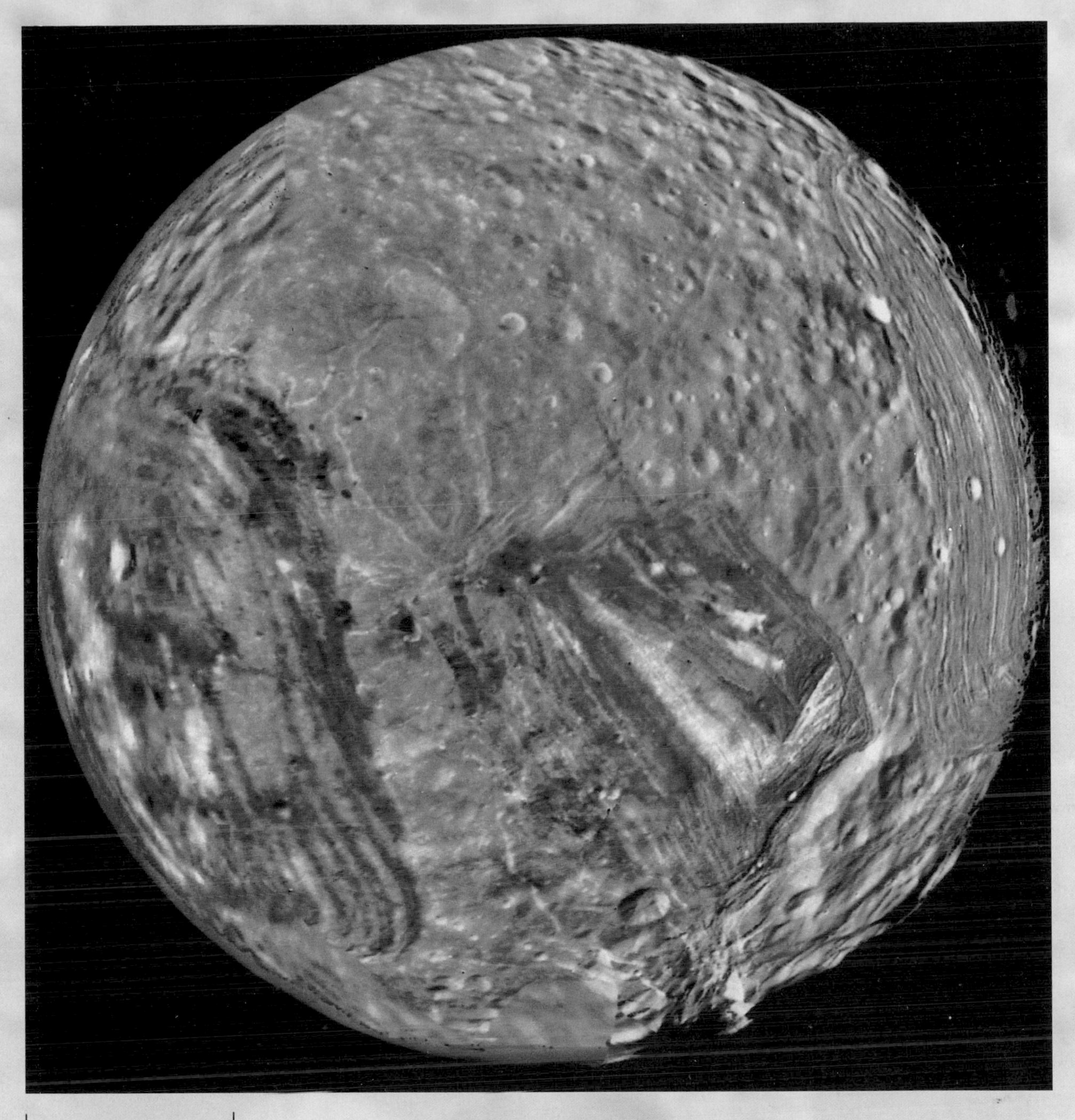

2.9 Exotic terrains on Miranda
One of the many surprises revealed by Voyager was the variety of geological histories shown by the satellites of Uranus. This is a mosaic constructed from the most detailed Voyager images of Miranda, a satellite of Uranus almost 500 kilometres in diameter. Despite Miranda's small size, it clearly has experienced a very complex history. This hemisphere includes three regions of ridged terrain (Arden Corona, left, Inverness Corona, centre, and Elsinore Corona, right) that appear to have formed within an older cratered terrain, by some form of icy volcanic process. Most of the craters in the old terrain have been smoothed off; perhaps they have been partly buried by 'snow' originating from explosive icy eruptions. Several much sharper craters, too young to have been affected by this, are superimposed on the cratered terrain and the ridged terrain. The youngest events of all are a series of fractures, which are especially prominent in the lower half of this view.

2.10 Airbrush map of Tethys
Images of parts of a planetary body recorded by any spacecraft are of variable quality, because of different resolution and varying illumination conditions. In order to produce a picture of a whole planetary body to a consistent standard, the United States Geological Survey employs skilled airbrush artists to draw the surface, using spacecraft pictures as a basis, but showing the solar illumination as if it came from a constant direction. This is one example of such a product, a view of the Saturn-facing hemisphere of its satellite Tethys, compiled from examination of images sent by the spaceprobes Voyager 1 and Voyager 2 in 1980 and 1981. Parts of the polar regions were not imaged, and have been left blank. Parts of the western and central regions of the globe were imaged at rather poor resolution, hence the rather noncommital and sketchy nature in those parts of this map, but elsewhere the resolution was good enough to show details as small as a couple of kilometres across. Much of Tethys has an evidently ancient, heavily cratered surface, but there are signs of more recent (though by now probably extinct) activity, notably the giant canyon Ithaca Chasma running from north to south through most of this hemisphere.

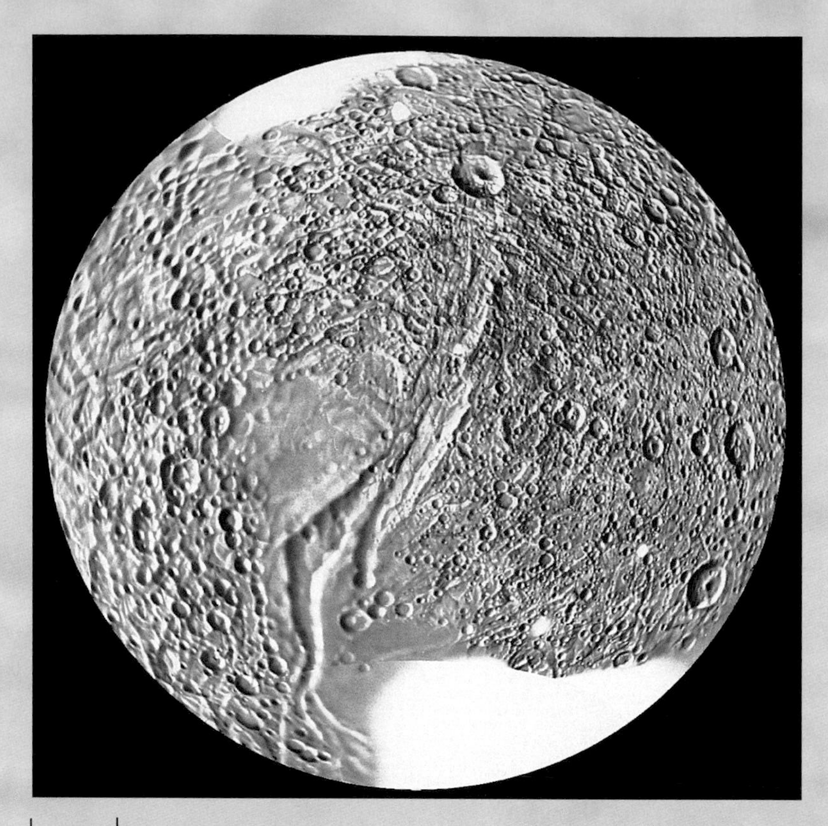

100 km

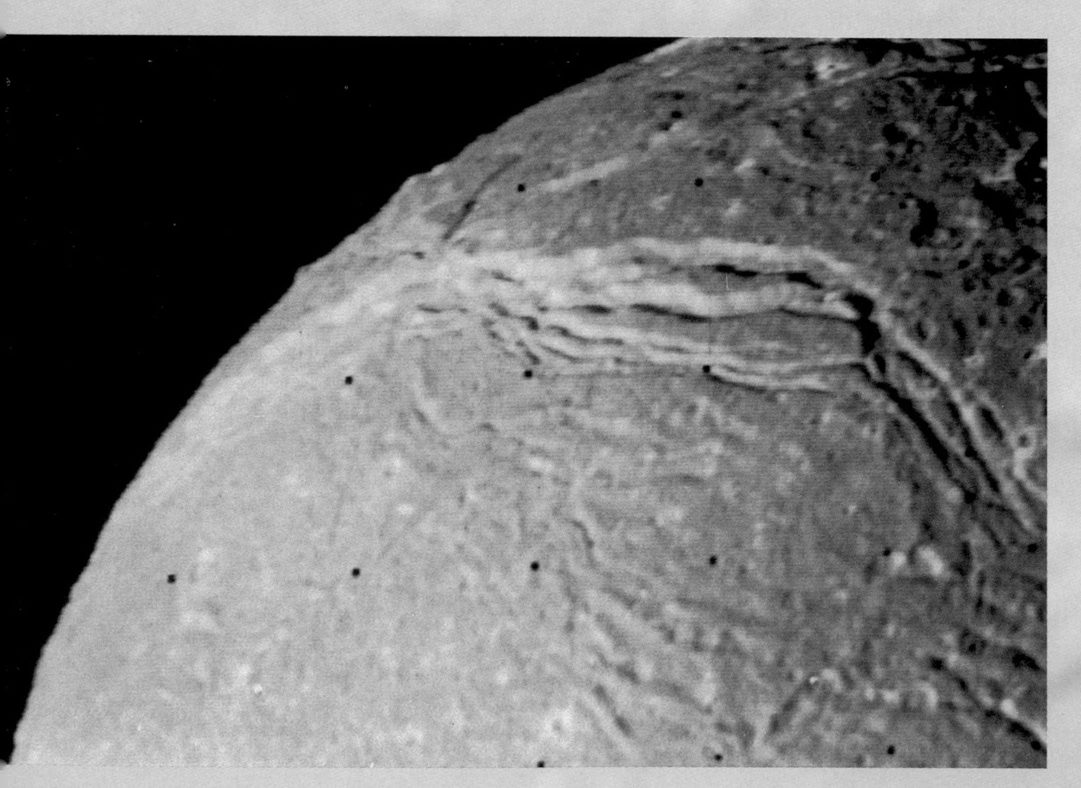

100 km

2.11 Ariel
This view of part of Uranus's satellite Ariel shows series of trough-like fractures that break up the impact-scarred surface. Many of these troughs have been subsequently filled by icy lavas. Ariel's geological activity probably wound down more than a thousand million years ago, but would have come to a halt much earlier if the feeble trickle of heat from radioactive decay of uranium, thorium and potassium in its small rocky core had not been supplemented by energy from tidal interactions similar to those that keep Io (Plates 2.26–2.28) active today. The dots appearing in a regular pattern across this picture are reference marks etched onto the faceplate of the Voyager camera, so that any distortion of the image during recording or transmission can be recognized and corrected. Image enhancement techniques have been used to remove these marks from the other Plates in this book.

2.12 Valles Marineris region of Mars
This is a mosaic of Viking Orbiter images, projected so as to show Mars as it would be seen from a point about 2500 kilometres above the surface, near the equator. This is too close to see a whole hemisphere, so neither of the polar caps is visible. The large gash across the centre of the view is the Valles Marineris canyon system, named in honour of the spaceprobe Mariner 9 that supplied the first detailed pictures of this remarkable area. Individual canyons are up to 8 kilometres deep and 200 kilometres wide. Though it may have been deepened by flowing water, the canyon system probably owes its origin to crustal fracturing over the arch of Mars's most extensive volcanic region, the Tharsis bulge, which lies at the system's western end. Three volcanoes (Arsia Mons, Pavonis Mons, Ascraeus Mons) are visible as dark red spots near the edge of the disc. The canyon system begins to their east, in a labyrinthine complex of intersecting fissures, which feed into a 2400 kilometres long series of parallel canyons. South of Valles Marineris lies some of the most heavily-cratered ancient terrain on Mars, about 3500 million years old.

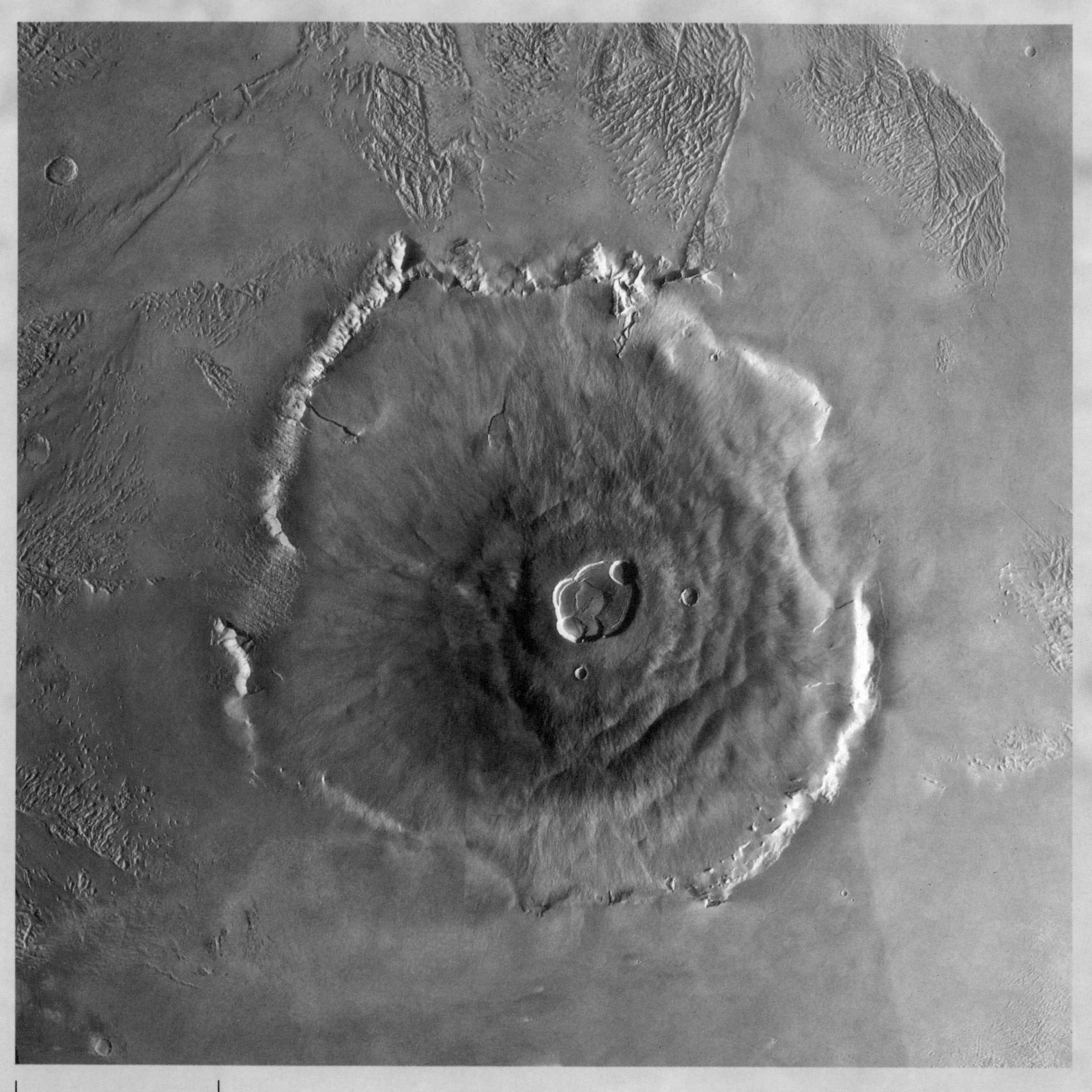

200 km

2.13 The largest volcano in the Solar System

This is Olympus Mons, the youngest volcano in Mars's most extensive volcanic region, Tharsis, and was probably last active over 100 million years ago. It lies just out of view to the west of the region shown on Plate 2.12. The overlapping craters in the summit region are up to 3 kilometres deep and are the result of a series of collapses following the drainage of magma that fed lava flows erupting from the flanks. The relatively gentle slopes of the volcano suggest that the lava was of the low-viscosity (i.e. particularly runny) type known as basalt. The island of Hawaii is similar in form, but the drop from summit to base (on the floor of the Pacific ocean) is only about 9 kilometres, whereas the summit of Olympus Mons is some 24 kilometres above its base. Mars is able to support such a large volcano because, being a smaller planet than the Earth, it has lost more of its heat, which has enabled its rigid outer layer (the lithosphere) to become thicker and stronger than the Earth's.

2.14 Detail of Martian canyon walls
A synthetic oblique view across Ophir Chasma (part of the Valles Marineris canyon system), which was made by combining colour images from the Viking orbiters with topographic data. This view indicates how the original form of the canyons has been modified. Each of the three curved bights in the cliffs at the back of the canyon marks the site of a major landslide, the debris from which has flowed across the canyon floor, possibly lubricated by ice or groundwater.

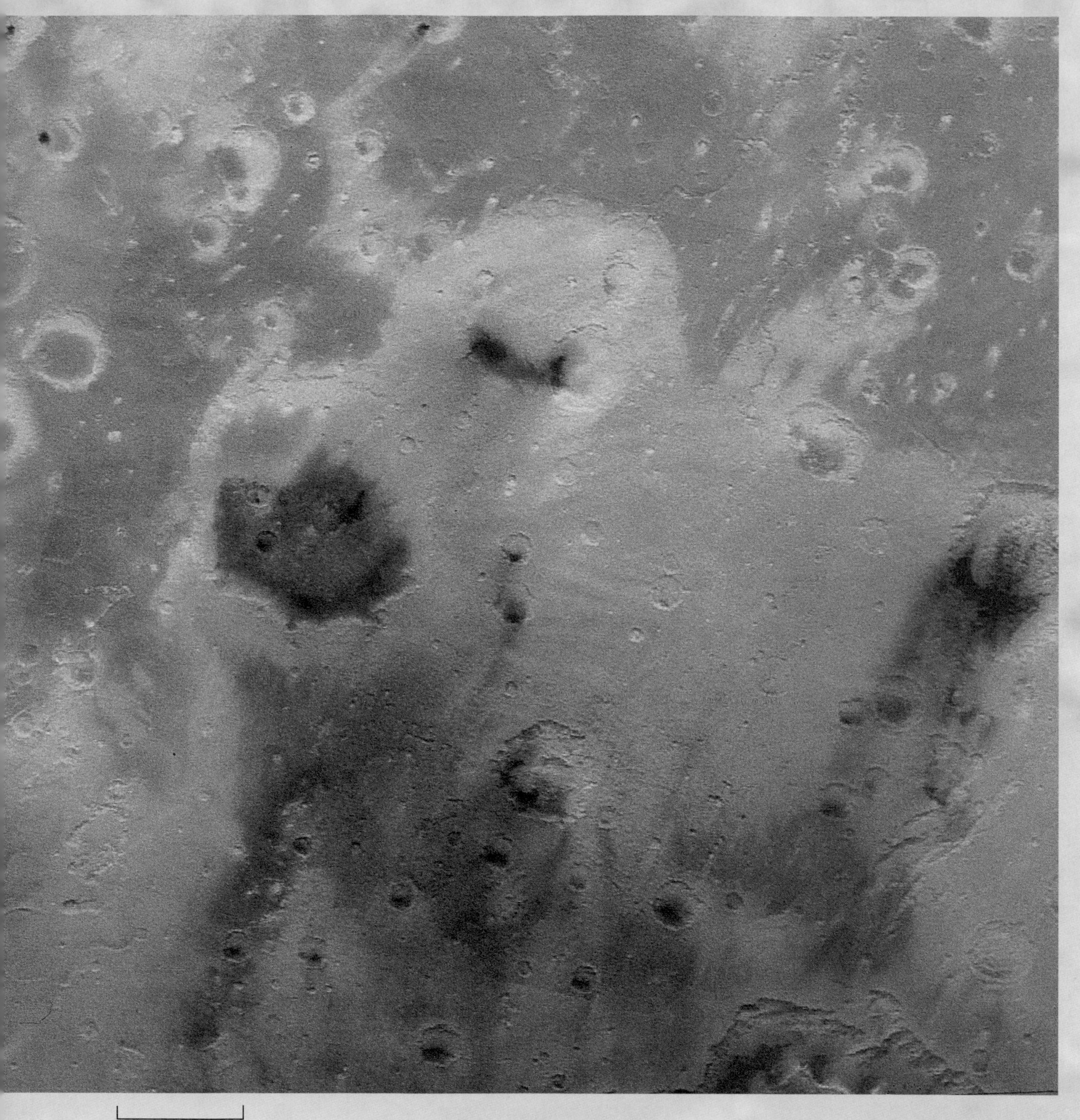

2.15 Xanthe Terra and channels on Mars
There are many sinuous channels on Mars, much smaller than the canyons of Valles Marineris (Plates 2.12 and 2.14), that seem to have been cut by flowing water in the distant past (maybe over 2 thousand million years ago) when Mars's atmosphere was denser than now. Several of these can just be picked out on this view of the region known as Xanthe Terra, which lies north of the eastern end of Valles Marineris, notably in the upper right where tributaries can be seen joining to form a larger channel. The two large craters with dark patches on their floors above left of the centre can be seen clearly near the right of the disc in Plate 2.12.

2.16 The surface of Mars
The Viking 1 lander touched down on Mars on 20 July 1976, a few hundred kilometres north of the area shown in Plate 2.15 in a region towards which many channels drain. It transmitted a series of pictures of the surface extending over more than a Martian year. This view looking southeastwards reveals a boulder-strewn dusty landscape, with some pale flattish surfaces in the middle-distance that may be bare rock outcrops. Parts of the lander itself are visible in the immediate foreground. The group of boulders near the left-hand edge is about 5 metres away, and the largest one is about half a metre long. The red colour is due to the presence of highly oxidized iron-rich materials (essentially rust). Dust particles in the atmosphere give the sky a pinkish hue. The dust is easily whipped up by the wind, and dust storms sometimes envelop the whole planet.

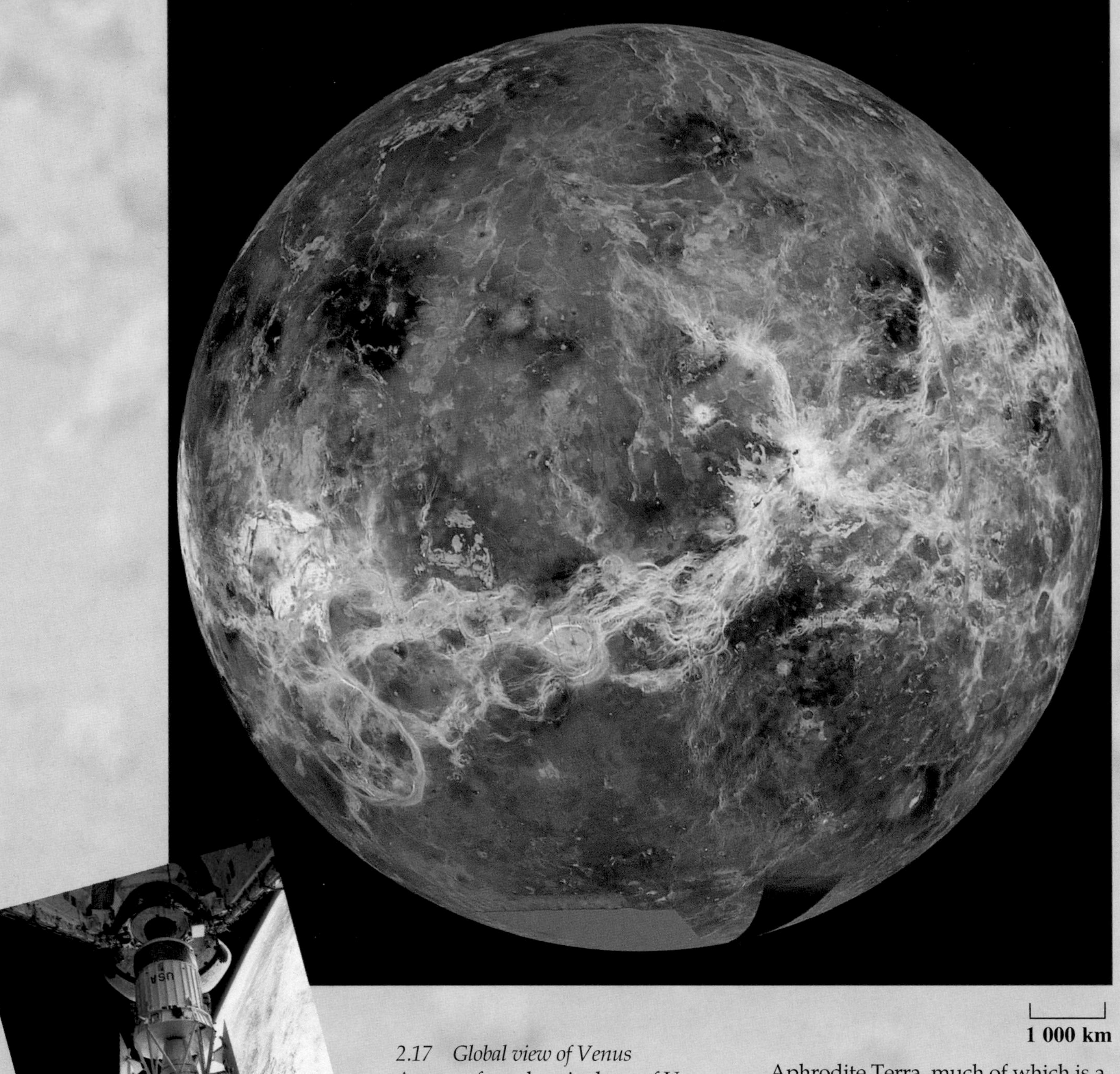

2.17 Global view of Venus

A map of one hemisphere of Venus made from data recorded by the imaging radar carried by Magellan, which went into orbit around Venus in August 1990. Microwave radar makes it possible to see details on the surface less than 1 kilometre across despite the otherwise opaque atmosphere (Plate 2.29). The bright-looking region extending across the globe just south of the equator is Aphrodite Terra, much of which is a quite highly-deformed terrain. Darker-looking regions are mostly volcanic plains, upon which there are many individual volanoes, too small to make out at this scale but shown clearly in Plates 2.18 and 2.19. The inset shows Magellan being deployed for independent flight from the cargo bay of a Space Shuttle in Earth-orbit.

2.18 Lava flows on Venus
Part of a very detailed Magellan radar image, showing a series of what appear to be quite viscous lava flows. Their morphology is reminiscent of the terrestrial silica-rich lava types known as dacite and rhyolite. Venus has an extremely wide range of volcanic landforms. Many, like this, appear fresh and are young enough that they do not bear the scars of impact craters. It has not yet been proved that volcanic activity continues today, but it is unlikely to have ceased.

2.19 Oblique view of Venus volcanoes A vertically-exaggerated oblique view showing the young volcanoes Sapas Mons (foreground) and Maat Mons (background), each of which is over 100 kilometres wide at the base. The gentle slopes of these volcanoes and the extensive lava flows emanating from them indicate that these are silica-poor volcanoes erupting rock of basaltic composition, which is consistent with crude chemical analyses made by five Soviet robotic landers on the surface. The original radar images are monochrome; yellow colour has been added to match the colour recorded by the landers. This perspective view was constructed in a computer, by superimposing Magellan radar images over a digital map of surface heights that had been assembled from data obtained by radar altimetry.

2.20 Oblique view across cratered terrain on Venus

A view across the Lavinia Planitia region of Venus constructed in the same way as Plate 2.19. Fractured terrain in the immediate foreground is overlain by an impact crater of 37 km diameter and its surrounding blanket of ejecta. This ejecta shows up bright on radar because its surface is rough, causing it to reflect the radar signal back towards the spacecraft strongly. Two more impact craters are visible in the distance.

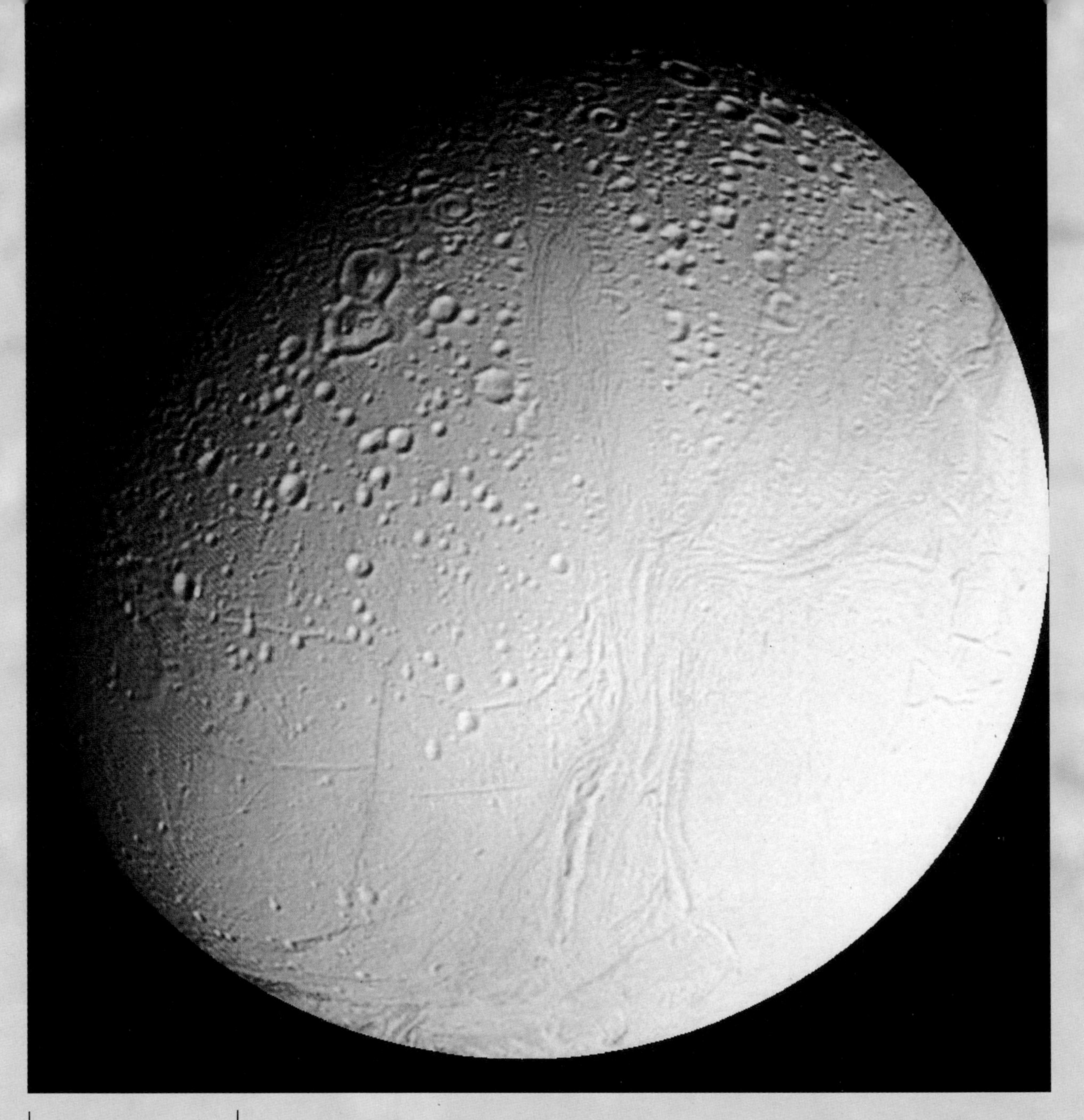

2.21 Global view of Enceladus
Enceladus is one of the smallest active worlds in the Solar System. It is an icy satellite of Saturn that, unlike its similarly sized neighbour Mimas (Plate 2.3) shows extensive evidence of recent, and probably continuing, geological activity on its surface. The region at the lower right is smooth and featureless, but is surrounded by belts of ridged material. Both of these terrain types are too young to have accumulated any impact craters that are big enough to show up. Many of the impact craters in the older terrains in the upper left have peculiar shapes, which suggests that they were formed in a weak or soft surface. Enceladus reflects a greater proportion of the incident sunlight than any other object in the Solar System, probably because its surface is covered by a sprinkling of fresh frost. This might be the result of explosive icy eruptions. The energy responsible for creating the young terrains and the fresh frost was probably provided by a past episode of tidal heating of the interior.

2.22 Global view of Europa
A high-resolution mosaic of Voyager 2 images showing part of this large satellite of Jupiter. Europa's high density indicates that most of its interior must be rocky, but it is covered by bright ice to a depth of probably several tens of kilometres. Heat from radioactive decay in Europa's rocky interior, supplemented by tidal heating, makes it likely that the ice is molten at depth. Certainly the surface is very young; less than a dozen impact craters show up on Voyager images, and the surface is broken by a variety of linear features. There are many cracks, filled in mostly by darker material, and overlying these in the lower part of the region shown here are curious 'cycloid ridges' rising about 200 metres above the general level of the terrain. The cracks and the brown mottled areas are evidence of recent (and probably continuing) icy volcanic activity and associated extension, whereas the cycloid ridges may indicate local compressive forces at work.

1 000 km

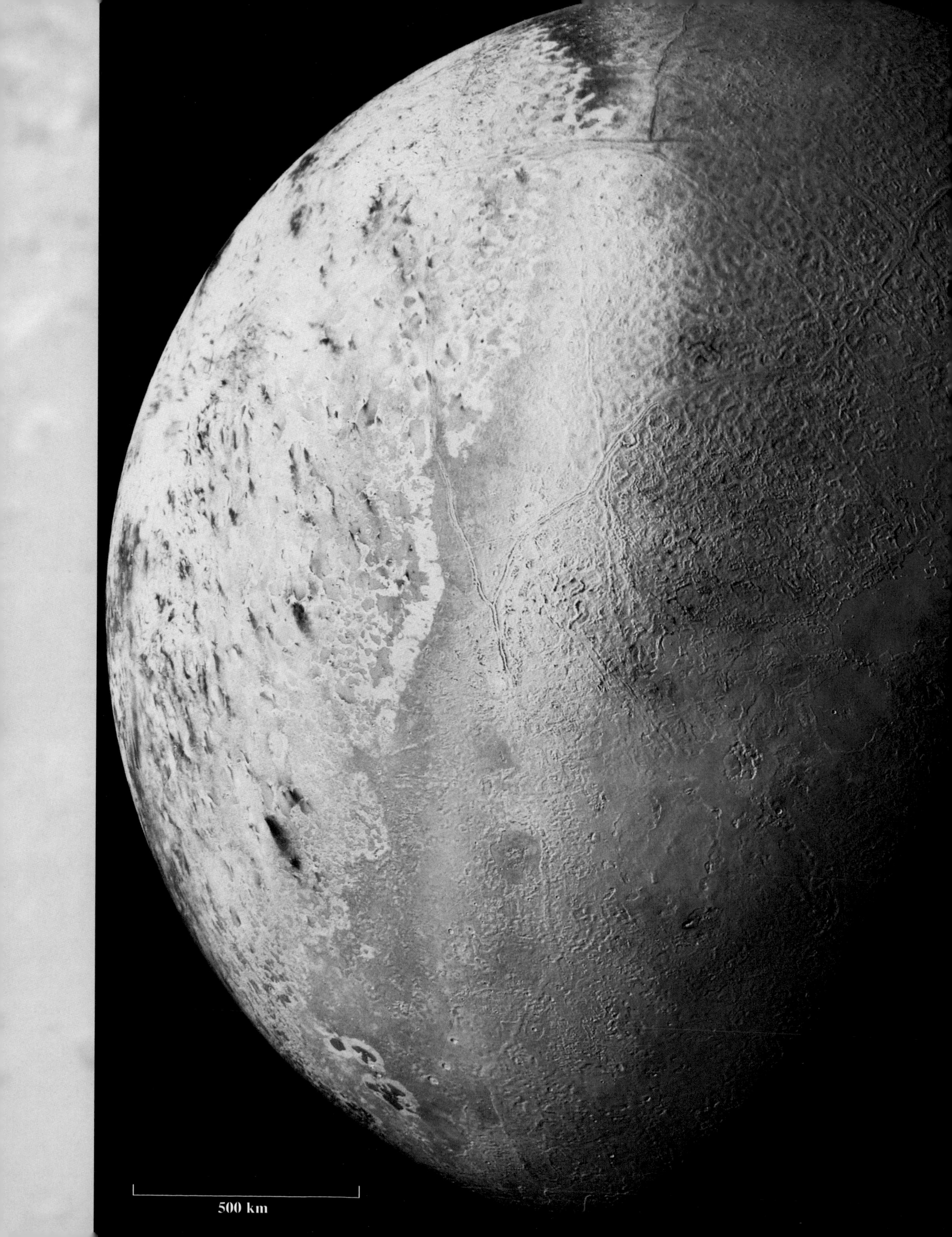
500 km

2.23 Global view of Triton
A mosaic made of the highest resolution images of the largest satellite of Neptune, obtained in August 1989 by Voyager 2 as it neared the end of its epic journey through the Solar System that had begun on Earth twelve years before. Triton's bright south polar cap (left) is of frozen nitrogen and is analogous to the Earth's polar caps of frozen water, and the frozen carbon dioxide caps on Mars. Several geyser-like plumes of unknown origin have been detected erupting from within the polar cap. These spurt columns of dust to heights of about 8 kilometres, and leave dark stains across the surface. Most of Triton's surface is of water-ice mixed with frozen methane and probably ammonia, and this mixed icy material behaves just as rock does on the Earth. The various ridges and dimples appear to be the result of icy volcanic processes (cryovolcanism), and cryovolcanic icy lavas have probably flooded the smooth areas towards the lower right.

2.24 Volcanic terrain in the Andes mountains
This is part of an image recorded by an Earth-orbiting satellite of the Landsat series, and shows an arid region in northwestern Argentina. Although made by combining reflected infrared light with visible light, the image has fairly natural-looking colours. The image is dominated by the oval volcanic caldera of Cerro Galan (near the centre, measuring 30 kilometres from north to south). This was formed in an explosive eruption about 2.2 million years ago, which distributed hot ash in a devastating flow that travelled over 100 kilometres from the vent. After removal of the magma the caldera was left as a collapsed depression, which is now partially filled by a volcanic dome that has grown near its centre. The dark patches near the western edge of the image are younger lava flows; these are mostly basaltic in composition, but one of them, about a fifth of the way up, is of a more viscous type and seems to be a small-scale version of the Venus lava flow shown in Plate 2.18. The small blue patches are salty lakes.

2.25 Fire-fountaining in Hawaii
Earth has many active volcanoes, which erupt silicate rock of a variety of compositions. This classic photograph from the vantage point of a helicopter shows red-hot molten basaltic rock propelled about 100 metres into the air by the force of escaping gases in the form of a 'fire fountain' at the Pu'u O'o vent of Kilauea volcano on the island of Hawaii. As it hits the ground the lava becomes channelled into flows, one of which is heading directly under the observing position.

2.26 *Typical terrain on Io*
A Voyager 1 image of a characteristic part of the surface of Io, Jupiter's innermost large satellite. Despite being only marginally more massive than the Earth's Moon, Io is the most active world in the Solar System. The craters scattered across this view are quite unlike the impact craters seen on most other planetary satellites, and in this case they are clearly a result of volcanic activity. Dark lava flows can be seen radiating from some of them, and others are wholly or partly infilled by black lava material. It seems likely that the volcanism on Io involves both molten silicate rock (as on the Earth) and molten sulphur. The red colour of the surface (which is exaggerated in this view) is due to sulphur and compounds of sulphur, but the white peak emerging from the plains at the lower left is evidently a rocky peak protruding through the sulphur-rich outer layers.

500 km

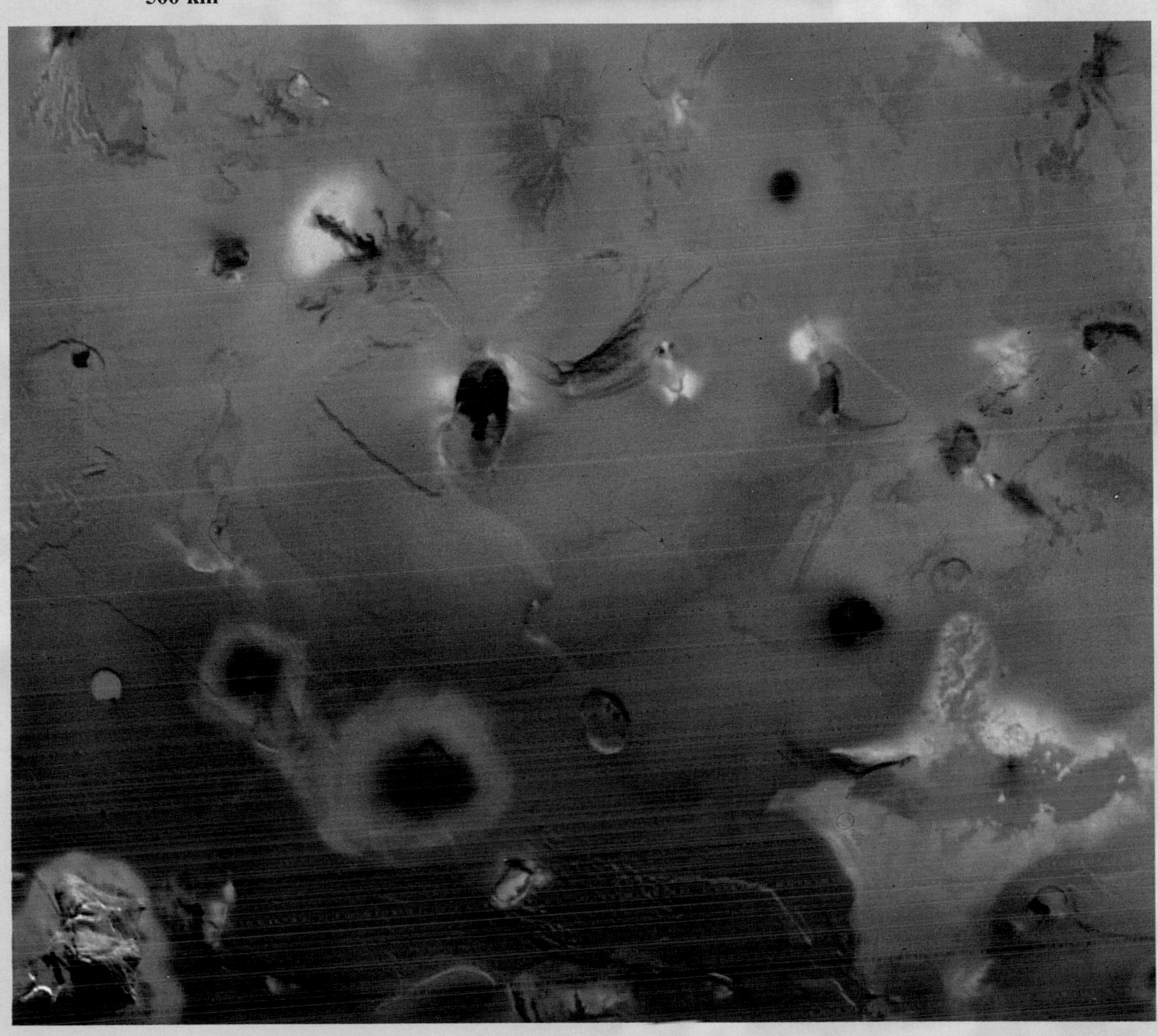

2.27 General view of Io
Another Voyager 1 image showing a surface mottled by the products of a variety of effusive (lava) and explosive (ash) volcanic eruptions. The dark patch near the lower edge is the volcano Pele, which was erupting at the time this image was recorded. We are looking down at the surface through the eruption plume, which consists of a mixture of sulphur dioxide and sulphur particles, and this is responsible for the diffuse oval ring surrounding the volcano. A different view of Pele is shown opposite.

2.28 Pele in eruption
Another Voyager 1 view of the erupting volcano Pele on Io. The vent is in the rugged area near the centre. The plume shows up clearly against the blackness of space, where specialized image enhancement techniques have been used to highlight it. There are several hotspots on Io's surface associated with volcanic activity; many appear to have temperatures of only a few hundred degrees and are usually attributed to the presence of molten sulphur at the surface. The record is held by the volcano Loki, where the temperature exceeded 1200 °C in January 1990, which can be due only to Earth-type lavas made of molten silicate rocks.

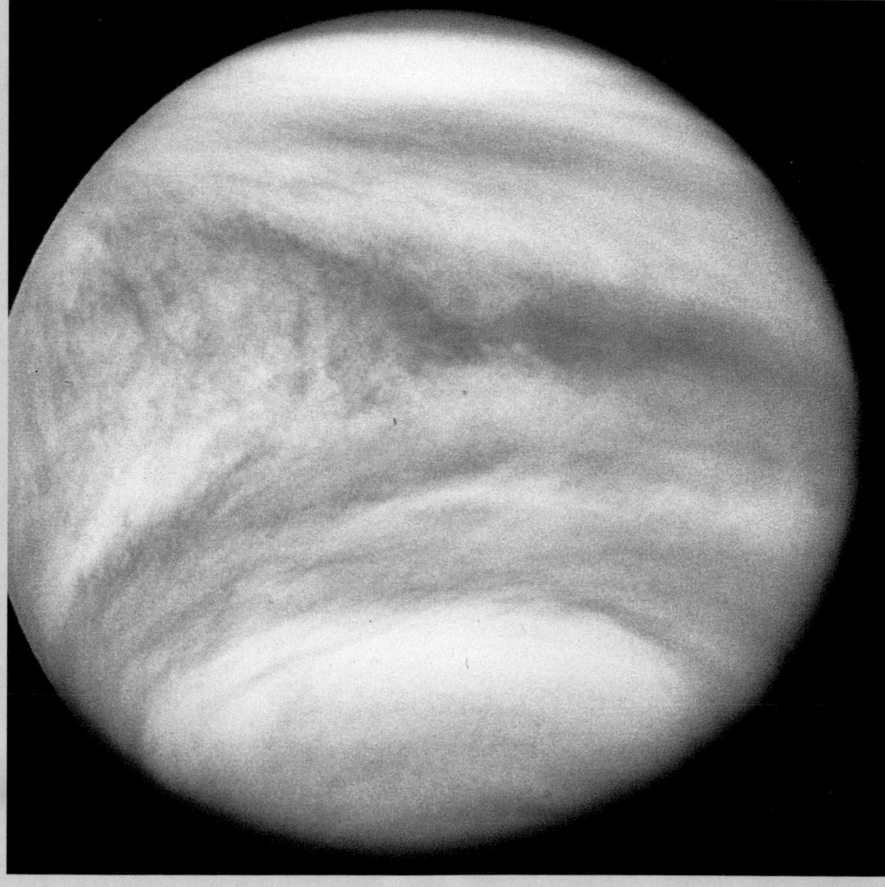

5 000 km

2.29 The clouds of Venus
Venus as seen by the spaceprobe Pioneer 12 on 10 and 11 February 1979. By recording an image in the ultraviolet it is possible to see details of the dense clouds enveloping the planet that do not show up in visible light. The clouds, which are composed of tiny droplets of sulphuric acid, cover the planet entirely. The pattern of the features indicates that a single convection cell (known as a Hadley cell) exists in each hemisphere, extending from the Equator to each pole (unlike Earth; see Plate 2.30). The cloud patterns change rapidly, as these images indicate, though the general form remains the same. In spite of the slow rotation of the planet (243 Earth-days, which is longer than the Venus-year of 225 Earth-days), the cloud layer rotates about the planet in only 4 Earth-days.

2.30 The Earth's atmosphere
A photograph by an Apollo 17 astronaut, en route for the Moon. The distinction between land and sea is clear; the continents appear reddish-brown and the oceans appear blue-black. The cloud cover, fortunately for us, is much less than for Venus. The pattern of clouds reflects the motion of the atmosphere. Within the tropics, convection cells (known as Hadley cells) rise near the Equator and sink at about 30° N and S. The near-surface flow converging on the Equator from Hadley cells on either side brings moisture that has evaporated off the warm tropical seas, and upon rising this condenses to form the cloud belt visible over equatorial Africa and to either side. At higher latitudes, spiral cloud patterns result from deflection of the atmospheric circulation system as a result of the Earth's rotation.

5 000 km

2.31 Clouds around Olympus Mons
Mars has an atmosphere that is much less dense than that of Venus, or even the Earth. However, clouds do develop from time to time, as shown here where Olympus Mons (see Plate 2.13) rises about 8 kilometres above a cloud deck of ice crystals. Like the Earth's atmosphere, that of Mars is often close to saturation by water vapour, so that the water has a tendency to condense when the temperature drops. Because of the low atmospheric pressure on Mars today, water condenses as ice and cannot exist as a liquid, though there is evidence of water on the surface in the distant past (Plate 2.15).

500 km

2.33 The atmosphere of Titan
This close-up looking across the limb of Saturn's largest satellite fails to reveal any details of the surface, but shows the atmospheric structure quite well. The orange opaque layer that hides the surface consists of suspended droplets about 200 kilometres above ground level, which are thought to consist mostly of hydrocarbons such as methane. The orange colour is imparted by traces of nitrogen-bearing compounds. Higher layers of bluish, hydrocarbon haze are visible above the opaque layer.

2.32 The Martian polar cap
An oblique view of the north polar cap of Mars, seen at mid-summer in the nothern hemisphere by the Viking 1 orbiter. This permanent part of the polar cap consists of water ice, and is surrounded by red wind-blown dust deposits, beyond which lie bluer-looking tracts of sand dunes. The curious spiral pattern within the polar cap is apparently picked out by a system of valleys that have been etched out by the wind. In winter the whole of this area is overlain by a seasonal cap of carbon dioxide, which condenses out of the atmosphere in the autumn and then dissipates in the spring.

2.34 Jupiter, Saturn, Uranus and Neptune

The four giant planets, shown to scale. If you were to fly past these in a spacecraft, all you would see would be their fuzzy outer cloud layers. The colours would be roughly as shown here, but the relative brightnesses have been adjusted. Being further from the Sun, Neptune receives only a fraction of the sunlight per unit area that falls on Jupiter, but this illumination difference has been compensated for in these images. The atmospheres of these planets are *extremely* deep, extending for thousands of kilometres below the tops of the clouds. The picture of Jupiter is from the Hubble Space Telescope in orbit around the Earth, the other images are from the Voyager spaceprobes.

100 000 km

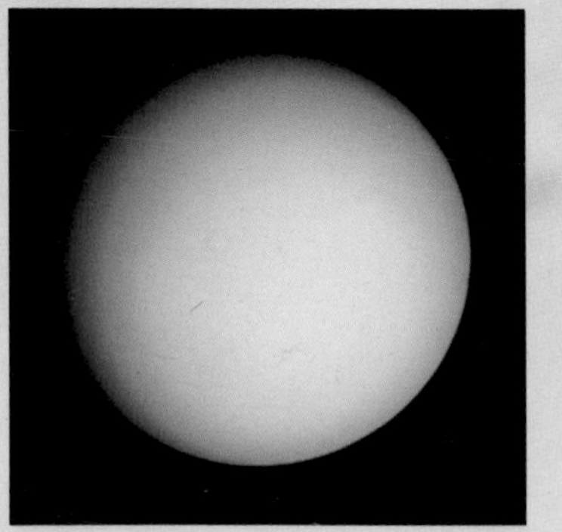

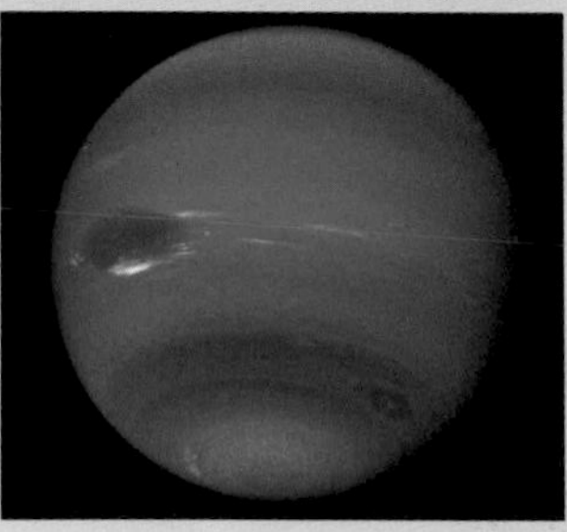

2.35 Atmospheric structure in Jupiter and Saturn

There is only a small range of colour within the atmosphere of each giant planet. By increasing the colour contrast on Voyager images, as here, the structure of the clouds can be made clearer. The image of Saturn has been enhanced more extremely in order to reveal the particularly subtle variations within it. The main characteristic of the cloud structure on both these planets is bands of cloud running parallel to the Equator. On Jupiter the pale bands are thought to be regions where the atmosphere is rising and the dark bands where it is sinking, though the system is complicated by the occurrence of giant eddies.

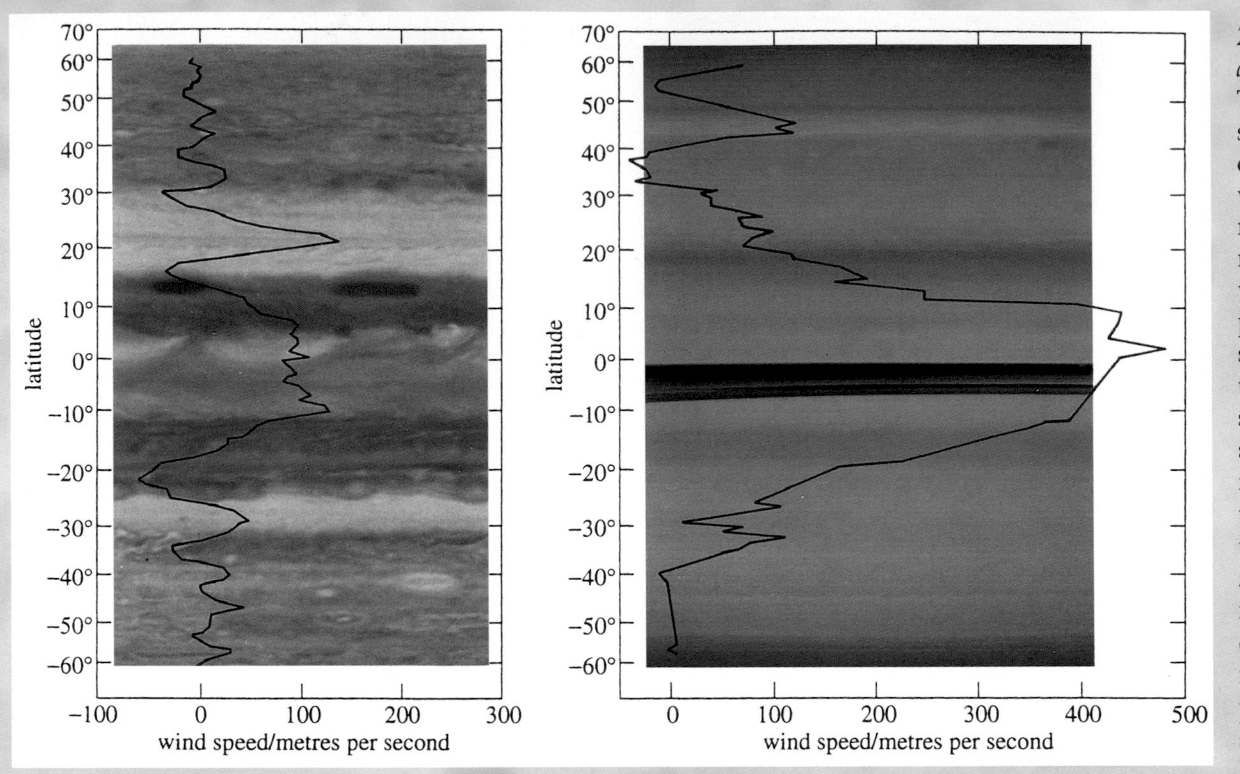

2.36 Wind speeds on Jupiter and Saturn

This shows how wind speed varies with latitude on Jupiter and Saturn. The wind speeds are measured relative to each planet's rotation speed. Negative velocities indicate atmosphere that is rotating more slowly than the interior of the planet, but still in the same direction. Comparison with the observed features on Jupiter shows that high wind speeds tend to correlate with the light zones. On Saturn there is not such a good correlation of wind speed with banding, and there is a zone of particularly high wind speed centred on the Equator.

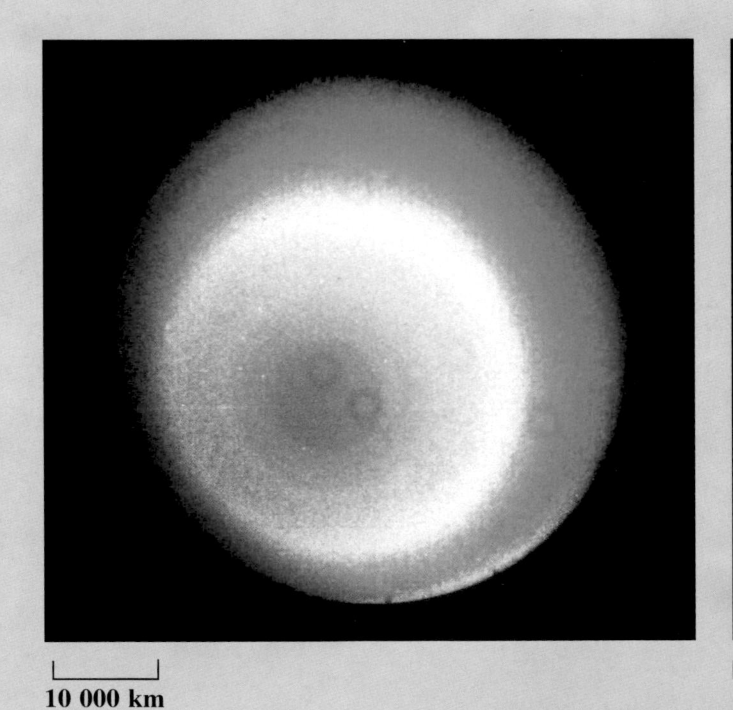

2.37 Exaggerated colour on Uranus

This is a view towards the south pole of Uranus, as seen by Voyager 2 as it approached the planet. The colours have been greatly exaggerated to show the structure of the atmosphere which, like those of the other giant planets, is banded parallel to the Equator. In this case the orange area over the pole demonstrates the presence of high-level haze. The bluer colours towards the Equator show the top of a lower cloud-deck, tinted blue as a result of absorption of sunlight by methane.

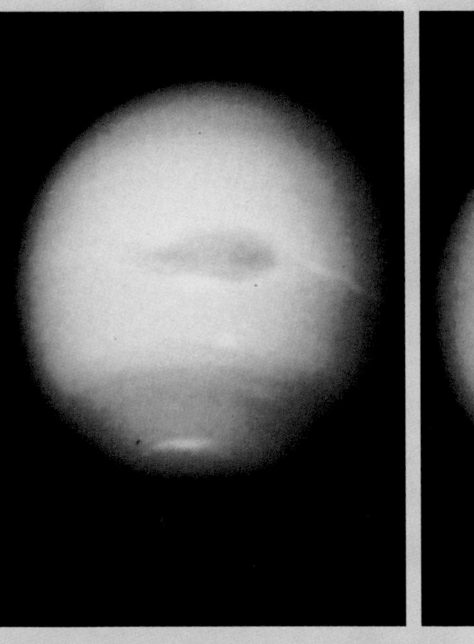
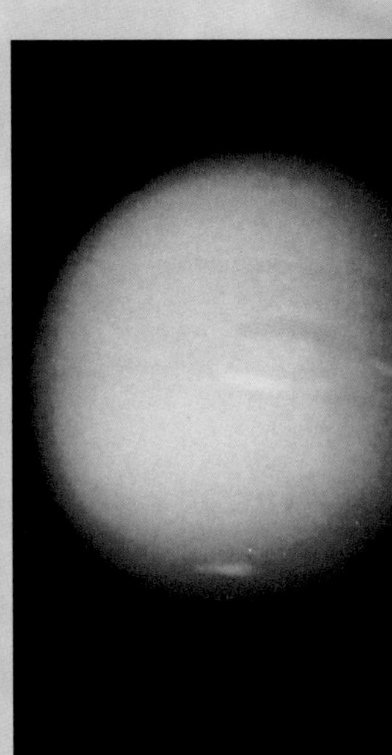

2.38 Neptune at different wavelengths

Neptune as seen by Voyager 2 through an orange filter (left) and a green filter (right). By examining images recorded through several different filters, it is possible to build up a picture of the locations of clouds of varying temperature and chemical composition.

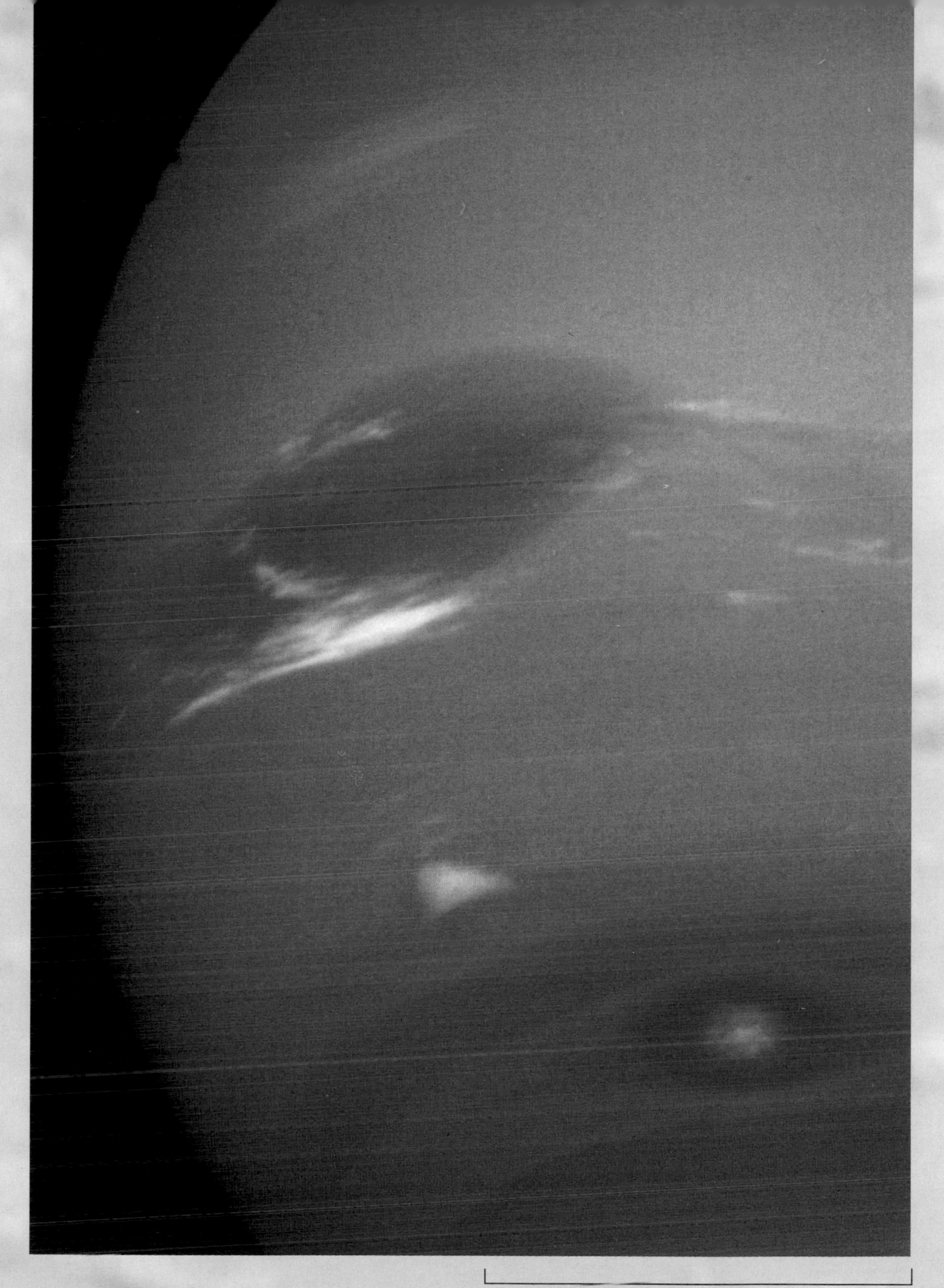

2.39 Cloud structures on Neptune
At the time of the Voyager 2 flyby, there seemed to be much more going on in Neptune's atmosphere than in that of Uranus. The large dark patch near the middle is known as the Great Dark Spot (an oval storm system comparable with Jupiter's Great Red Spot on Plate 2.40). The white cloud at its southern edge is a fast-moving feature nicknamed the Scooter, which actually lies deeper in the atmosphere than the less prominent pale clouds to its right. Farther south is a smaller dark spot with a bright core, which is driven by fast-blowing eastward winds to overtake the Great Dark Spot every six days.

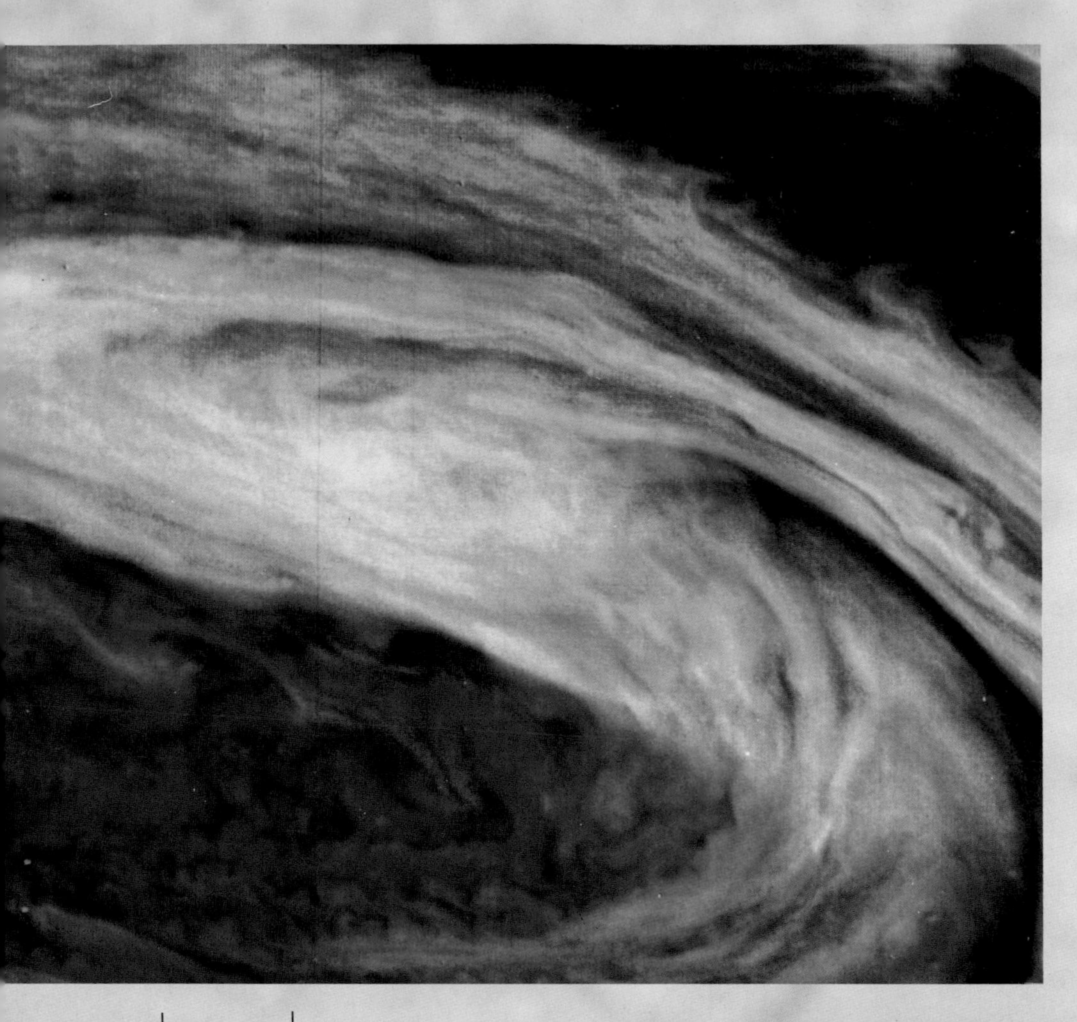

1 000 km

2.40 The Great Red Spot of Jupiter This giant storm has been seen through telescopes for at least 160 years. This picture shows a close-up of its northern edge as seen by Voyager 1 in extremely exaggerated colour. Far from being a simple structure it is a large anticyclonic storm with numerous eddies and smaller storms surrounding it and within it. The Great Red Spot is also prominent (in less exaggated colour) in the view of Jupiter shown in Plate 2.35.

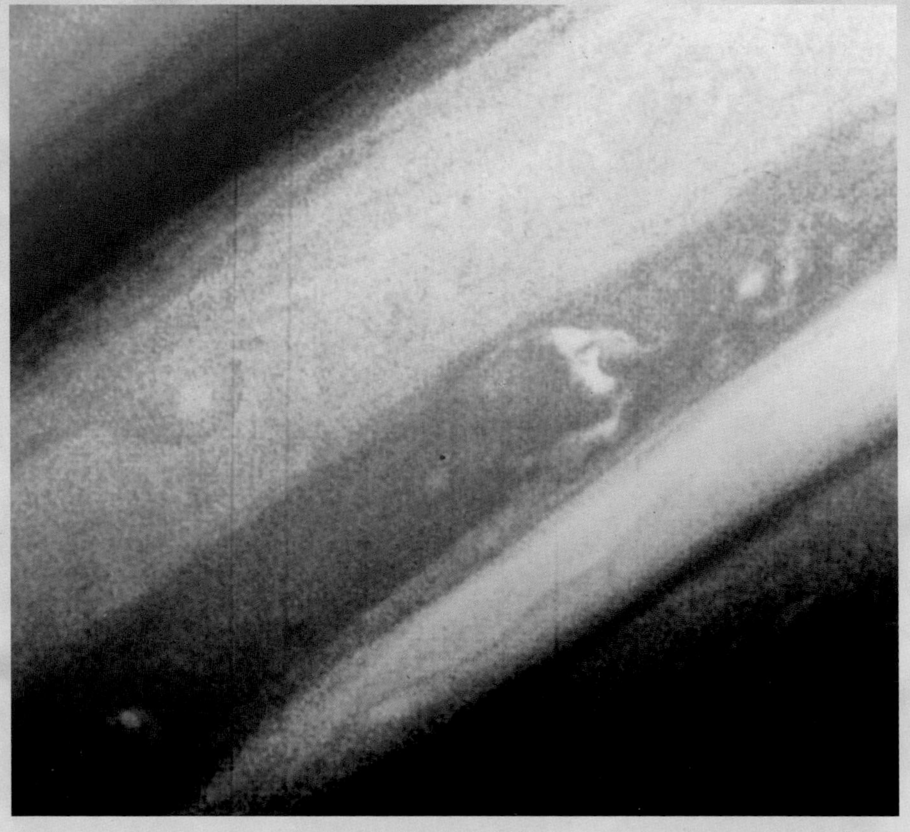

1 000 km

2.41 Atmospheric patterns on Saturn Saturn generally has a blander appearance than Jupiter, and there is no Saturnian equivalent of the Great Red Spot, so in natural colour it can be difficult to pick out features in its clouds. However, enhanced images such as this view from Voyager 1 in 1980 reveal the presence of storms resembling those on Jupiter, though on a smaller scale.

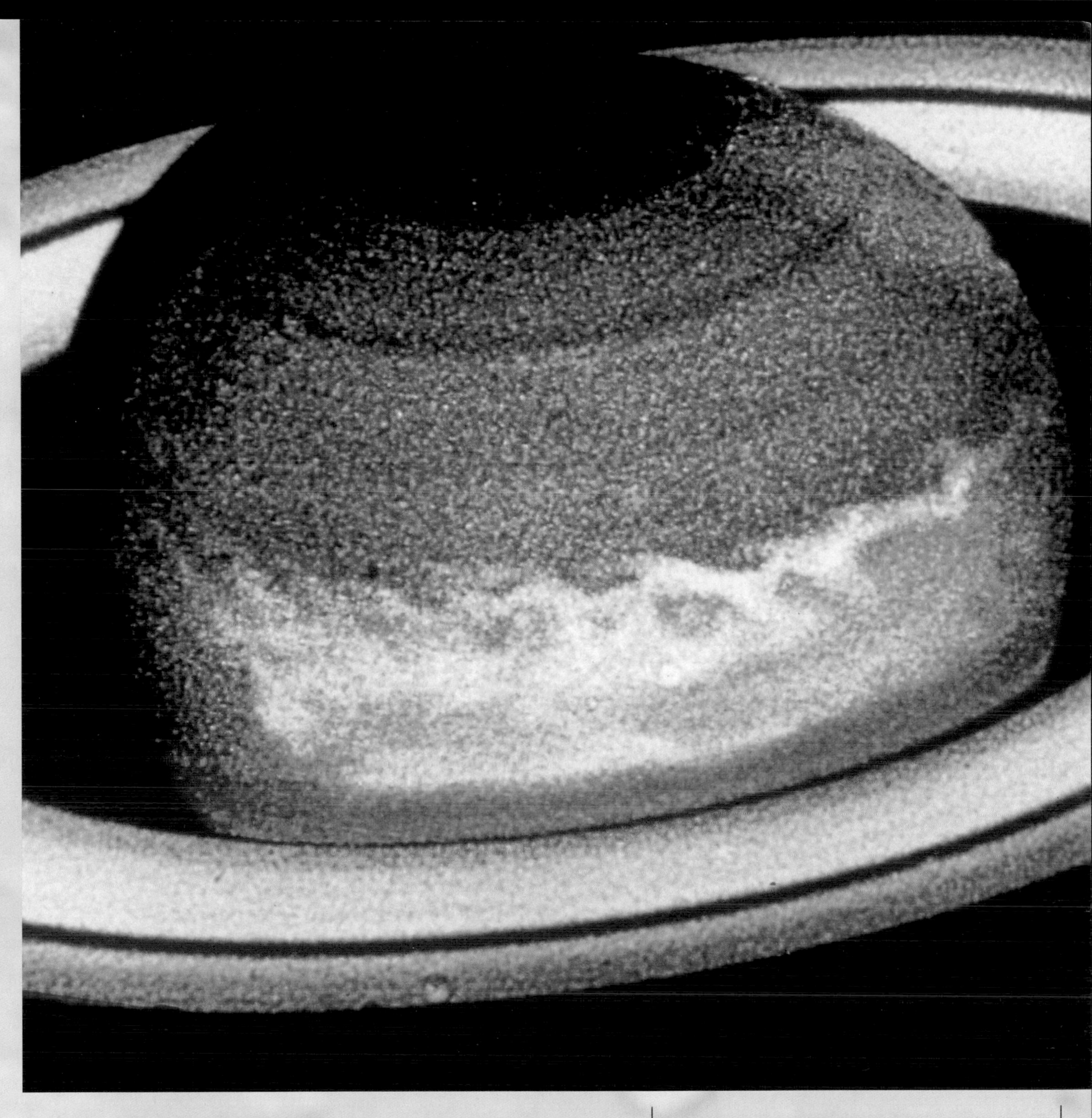

2.42 A giant storm on Saturn
In September 1990 ground-based amateur astronomers noticed a white spot on Saturn. Although a similar feature (named the Great White Spot) had appeared previously (notably in 1933 and 1946), it was not there when the Voyager spacecraft flew past Saturn in 1980 and 1981. By the time the picture shown here was recorded by the Hubble Space Telescope, on 9 November 1990, the spot had spread out into a series of linked vortices. The circulation pattern of the atmosphere shows particularly well in this rendering, which is a combination of an image recorded in blue shown as blue and an image recorded in near-infrared shown as red.

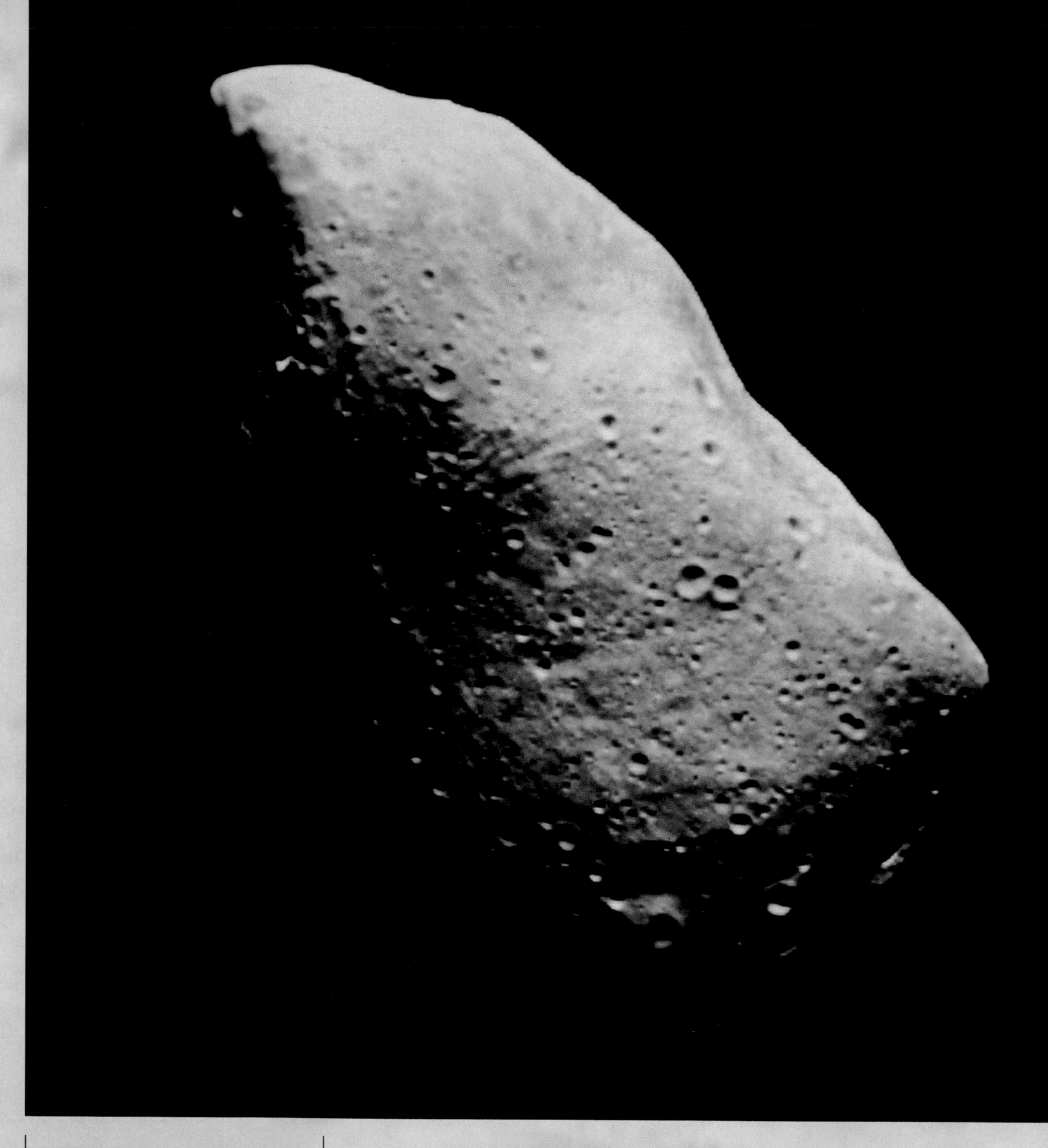

2.43 The asteroid Gaspra
The first-ever close-up of an object in the asteroid belt, the asteroid 951 Gaspra, as seen in exaggerated colour by the spaceprobe *Galileo* in October 1991 on its way to Jupiter. Craters as small as 100 m across are revealed in this image. There are rather more small craters relative to larger craters than would be expected if Gaspra had existed since the birth of the Solar System, so it seems that Gaspra was formed by the comparatively recent collisional breakup of a larger body. Its irregular shape is probably a result of this collision. Observations of its reflectance characteristics made by ground-based telescopes suggest that Gaspra has a composition similar to stony chondritic meteorites (e.g. Plate 2.47b), but as Galileo flew by it detected a distortion in the interplanetary magnetic field of the sort likely to be due to the presence of a substantial proportion of metallic iron within Gaspra, so in fact its composition may be more like that of stony-iron meteorites (e.g. Plate 2.47d).

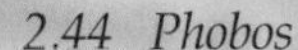

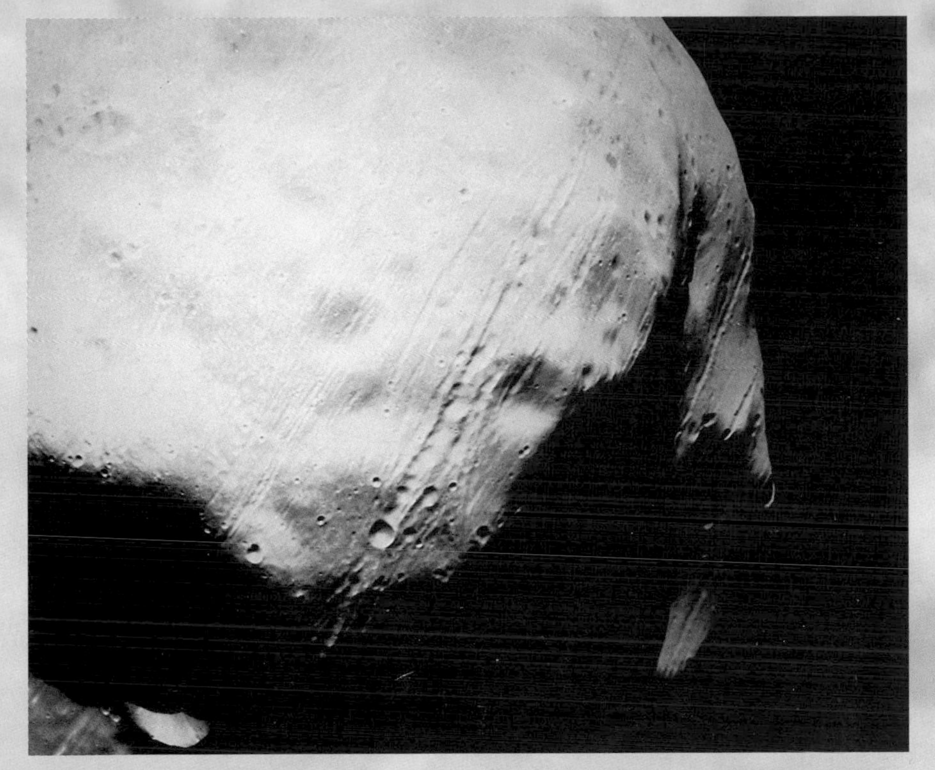

10 km

2.44 Phobos
Phobos is the larger of Mars's two tiny satellites, both of which are almost certainly captured asteroids. Like any ancient surface in the Solar System, the surface of Phobos is scarred by craters formed by the impact of smaller bodies. There are also parallel grooves of uncertain origin: if these are fractures they could be due to tidal forces when Phobos was captured by Mars or stresses generated during impacts, or they could simply be 'machine-gun bullet' traces formed on the surface as Phobos travelled through strings of debris ejected into space by large crater-forming impacts on Mars. Phobos's colour, as documented by ground-based telescopes, suggests that it is composed of the same material as carbonaceous chondrite meteorites that have been collected on Earth (e.g. Plates 2.47c and 2.48a).

10 km

2.45 The nucleus of Halley's comet
A close-up view as seen on 13 March 1986 by Giotto (see inset), which was the first spacecraft to make a close encounter with a comet's nucleus and send back detailed images. This picture was made from seven images recorded at distances ranging from about 25 thousand to less than 4 thousand kilometres away, combined so as to give the sharpest possible view. The potato-like shape of the nucleus can easily be made out, illuminated by sunlight coming from the upper left. Jets of gas can be seen streaming off in a sunward direction, and the depressions in the surface of the nucleus are probably the result of collapse, as volatile substances within the nucleus have been vaporized by heat from the Sun.

(a)

(b)

(c)

2.46 Meteorite collection
Meteorites are fragments of extraterrestrial material that fall to the Earth's surface. Occasionally members of the public are lucky enough to notice their arrival. (a) Mr Arthur Pettifor holds a meteorite that struck a tree in his garden in Cambridgeshire, UK, in May 1991. It is now in the British Museum. Because meteorites are so rarely observed to fall, and because they weather away like ordinary rocks once they are lying on the ground, expeditions are organized to search for them in environments where they are likely to survive. (b) A meteorite being collected in Antarctica. It is being put in an ultra-clean Teflon bag using stainless steel tongs in order to minimize contamination and maximize the scientific return from detailed chemical analysis of the sample. (c) A much larger chunk found in the Australian outback has been loaded onto a truck.

2.47 Different sorts of meteorite

There are many sorts of meteorite. With a few rare exceptions that are impact-ejected fragments of rock from Mars (Plate 2.50) and the Moon, they seem to correspond to the wide range of types of asteroid that have been observed, and are believed to be fragments of asteroids broken up by collisions. Four types are shown here, each of them a few centimetres across. (a) A cut and polished surface of an iron meteorite that has been etched with acid to reveal the characteristic pattern of an alloy of iron and nickel that has cooled extremely slowly, presumably within the core of an asteroid. (b) A stony meteorite of basaltic composition formed by melting processes within its parent body, just like igneous rocks on Earth. (c) A carbonaceous chondrite, representing some of the most primitive material in the Solar System. It is a friable mixture of rocky dust and more volatile material (including organic molecules) believed to have grown by the collision and sticking together of grains that had condensed within the solar nebula (Plate 2.1). (d) A stony-iron meteorite, or pallasite, that may come from near the boundary of an iron-rich core and silicate-rich mantle of a large asteroid.

(a)

(b)

(c)

(d)

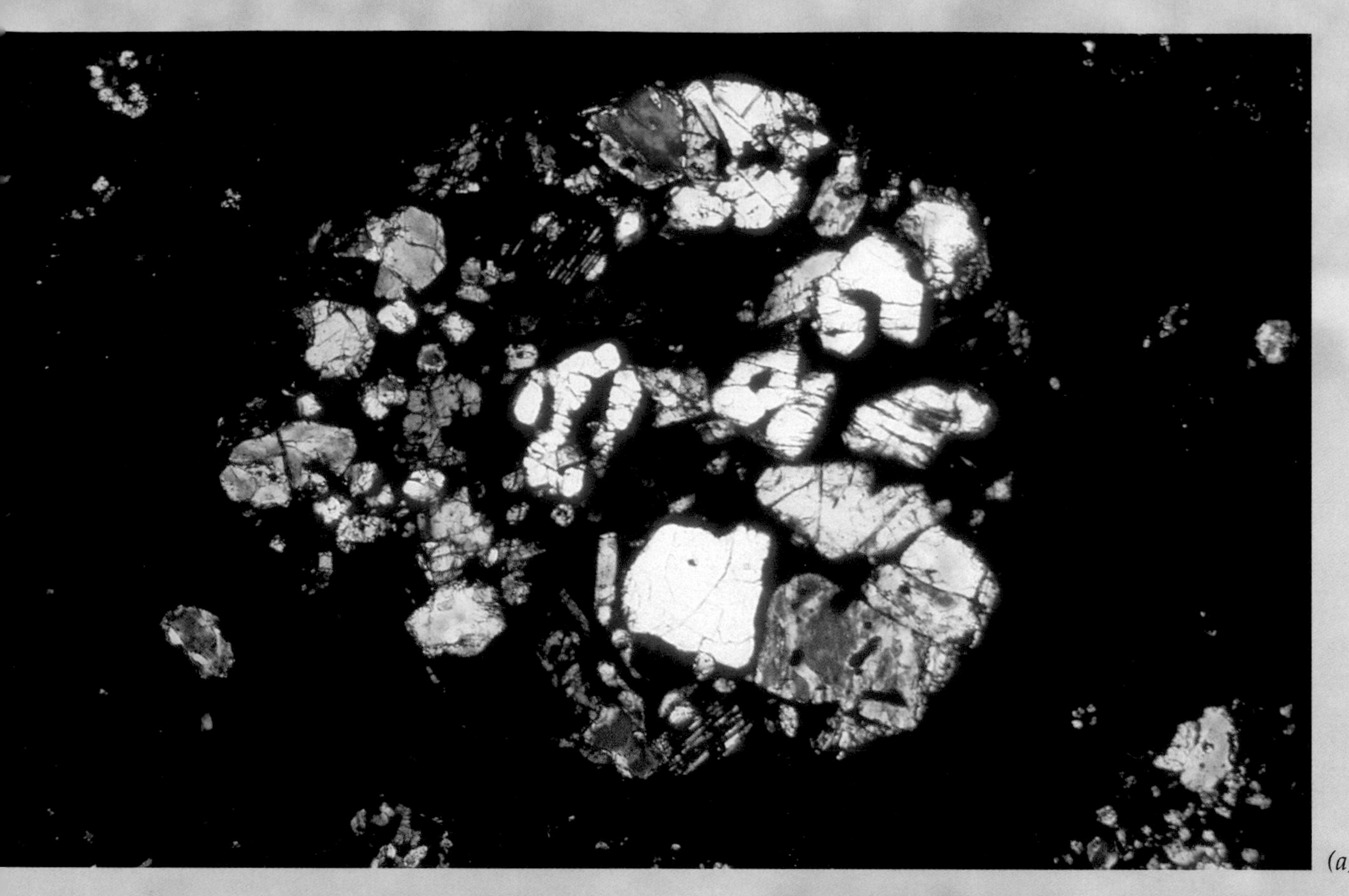

(a)

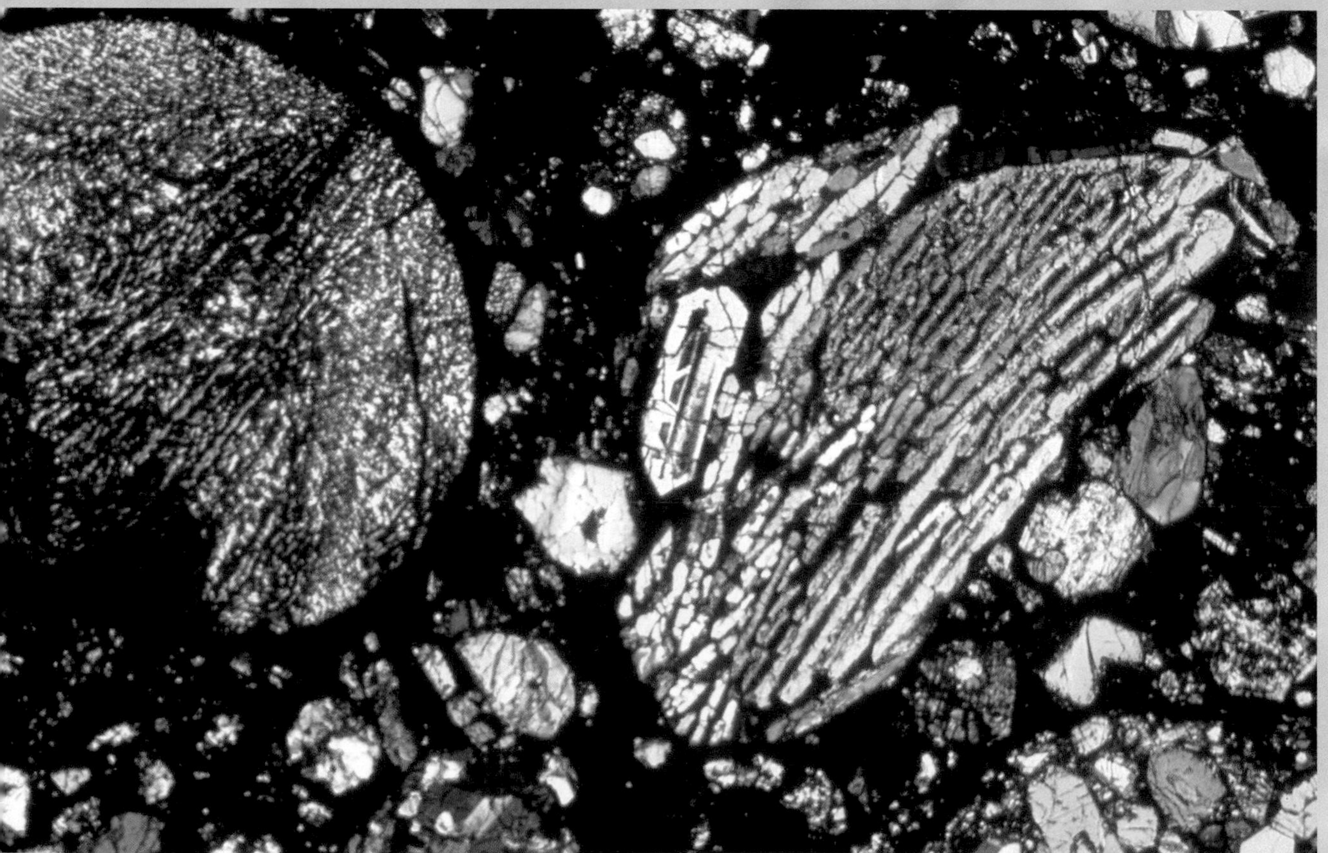

(b)

1 millimetre

(c)

1 millimetre

2.48 Meteorites under the microscope
In order to examine the mineralogy and crystal texture of meteorites (or terrestrial rocks for that matter) it is customary to make a slice of the sample that is thin enough to be transparent (typically only 0.03 millimetres thick) and to examine it in polarized light under a microscope. This shows different minerals in characteristic colours, depending on their orientation. Three examples of stony meteorites are shown here. (a) The Cold Bokkeveld carbonaceous chondrite, showing a rounded collection of crystals, known as a chondrule, which formed at high temperature, embedded in a dark matrix of silicate dust rich in water and organic compounds that formed at low temperatures. These two components may have accumulated together as a sediment on the surface of an asteroid parent body. (b) A chondrule and fragments of a second embedded within crystal-rich rock in the Sharps meteorite, which is an ordinary chondrite. The chondrules are older than the crystals that surround them, and may represent globules that formed within the solar nebula. (c) The Stannern meteorite, which lacks chondrules and is evidently a product of melting in the parent body.

2.49 Cosmic spherules
These particles were collected from a depth of about 5 kilometres on the floor of the Pacific Ocean. Their chemical composition is demonstrably extraterrestrial. Such spherules form from cosmic dust particles of millimetre dimensions that melt as a result of friction as they are decelerated by the Earth's atmosphere. Cosmic debris such as this rains down over the entire surface of the Earth and would build up to a depth of about 1 millimetre within about ten million years were it to lie undisturbed. In reality, only certain areas in deep oceans are calm enough for these spherules to accumulate in easily collectible amounts.

2.50 A meteorite from Mars
This 8 kilogram meteorite was collected in Antartica in 1979 and is referred to by its classification mark as EET A79001. It is a chunk of basalt lava, formed by a melting process, and its age is very young for a meteorite, about 200 million years, compared with about 4550 million years for most meteorites. The young age makes it unlikely that it formed anywhere other than on a planet, and the details of the chemical composition of the rock itself and gas trapped within it match what we would expect to find on Mars. The dark skin, which has partly flaked off, is a result of heating caused by friction on entry into the Earth's atmosphere.

2.51 The rings of Saturn
A detailed view of Saturn's ring system as seen by Voyager 1. The rings consist of icy debris orbiting in the plane of Saturn's equator. They are probably the remains of a broken-up satellite, and most chunks are only a few centimetres in size. Brightness variations between rings correspond to differences in the size and abundance of ring particles. The gaps in the rings, only the largest of which can be seen using Earth-based telescopes, occur where the orbital period would be a simple multiple of the orbital period of one of Saturn's many satellites, and are swept clear of debris by repeated gravitational interactions. When seen in more detail, the outermost of the rings in this view is revealed to be a contorted tangle of narrow strands, kept in place by two small satellites of Saturn (Janus and Epimetheus), each less than 200 kilometres in diameter, whose orbits lie very close on either side of the ring. The other giant planets (Jupiter, Uranus and Neptune) also have systems of rings, though far less spectacular than these.

Part 3

Galaxies and the Universe

The Sun and its planets are just a tiny part of a vast astronomical structure known as the Milky Way. The Milky Way is a galaxy. It is our galaxy, the one in which we live. Everything that the unaided eye can see in the night sky belongs to the Milky Way, apart from three or four faint misty patches that betray the existence of other galaxies beyond the boundaries of our own.

Powerful telescopes reveal that the few galaxies that are visible to the naked eye are just the tip of a cosmic iceberg. The number of observable galaxies beyond the Milky Way is enormous, probably greater than the number of observable stars within the Milky Way. A patch the size of the full Moon typically covers many tens of thousands of galaxies. We are usually unaware of these galaxies simply because they are too faint, but they contain most of the visible matter in the Universe.

Just like stars and planets, galaxies are individuals that differ from one another in a variety of ways. The smallest have fewer than a million stars and are less than a thousand light years in diameter. The largest probably contain millions of millions of stars and are hundreds of thousands of light years across. These differences are also reflected in other galactic properties, such as gas content and shape. Both the smallest dwarf galaxies and the largest giant galaxies are elliptical and rather featureless, whereas many of the intermediate-sized galaxies, including the Milky Way, are flattened discs with central bulges and prominent spiral arms. The shapes of galaxies are usually described in terms of a classification scheme (below) devised by the American astronomer Edwin P. Hubble. Within this scheme, the form of a galaxy – elliptical or spiral, barred or unbarred, lens-shaped or irregular – can be represented by just a few letters and numbers. Hubble's scheme will be used frequently to describe the images that follow. In the past, it has been suggested that the Hubble classification scheme is an evolutionary one, with galaxies gradually changing from elliptical to spiral, but this is no longer thought to be the case. Unless they are involved in collisions or mergers, galaxies probably keep the same shape for immensely long periods of time.

Another way in which galaxies are distinguished from one another is by their names or catalogue designations. Relatively few galaxies have been given individual names, so it is quite common for astronomers to refer to galaxies by their code numbers in various catalogues. The best known of these is probably the list of 110 non-stellar objects compiled by the French comet hunter Charles Messier, and many of the galaxies that follow will be referred to by their Messier

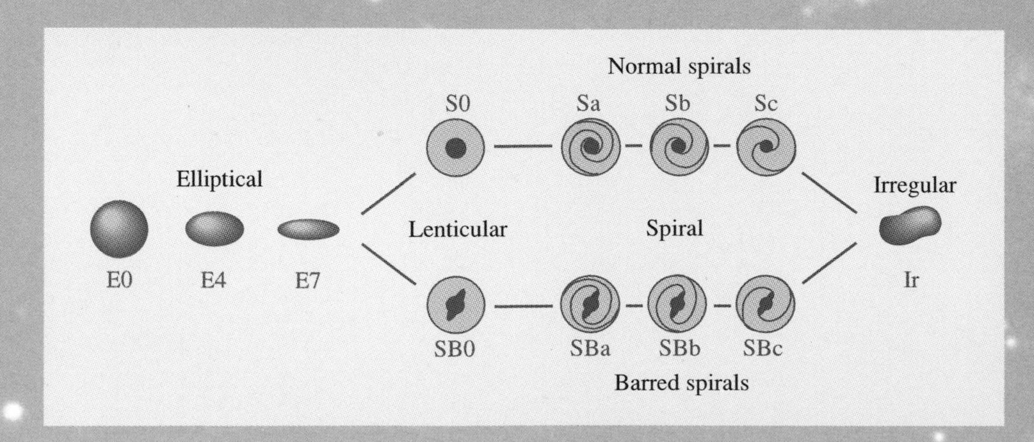

The Hubble scheme for the classification of galaxies

numbers, such as M31 and M32. Other, larger catalogues that are more frequently used by professional astronomers include the New General Catalogue and its extension, the Index Catalogue. You will meet many galaxies that are identified only by their NGC and IC numbers, together with a few that are better known by their identity in more specialized catalogues, such as Halton C. Arp's listing of peculiar (distorted) galaxies. Of course, when all else fails, it is always possible to identify a galaxy by means of its co-ordinates on the sky, and this method too will be used in a few cases. Most of the scale bars in Part 3 are given angular rather than linear values, because of the uncertainties in the distances of the galaxies.

In this last and most wide-ranging part of your journey through the Universe, you will be taken to the observable limits of time and space. The trip starts close to home with an investigation of the Milky Way. Once you have familiarized yourself with our own galaxy, you will travel beyond its boundaries to examine some of the relatively nearby galaxies that can be most clearly imaged. Following this you will see some of the more rare and distant galaxies, including representatives of the class of highly energetic active galaxies, of which the enigmatic quasars are the most illustrious examples.

Galaxies are the building blocks of the Universe. On the final leg of the journey you will survey regions of space so vast that individual galaxies will be lost from sight and only the shapes of the clusters and superclusters of galaxies that make up the large-scale fabric of the Universe can be glimpsed. Having reached this scale you will be taken back in time to ponder the birth of the Universe and the origin of galaxies in the fiery depths of the Big Bang. Fasten your seat-belt, it's going to be quite a trip.

3.1 The spiral galaxy NGC 2997
This faint galaxy in the constellation of Antlia lies far beyond the boundaries of the Milky Way, at a distance of about 45 million light years. Nonetheless, it provides a good idea of what our own galaxy might look like from a similar distance. The prominent spiral arms are just an unusually bright part of a fairly uniform disc of stars and gas. The disc looks rather oval in the image, but that's simply because its plane is inclined at about 40° to the line of sight, so there is a certain amount of foreshortening. Compared with our own galaxy, the spiral arms of NGC 2997 are perhaps somewhat more open, and the bright central bulge somewhat smaller.

3.2 The Milky Way: an artist's impression

The Milky Way, shown here in plan view and side view, is thought to be a spiral galaxy with a prominent central bulge. It is composed of stars, gas and dust, threaded by magnetic fields and permeated by high-energy particles known as cosmic rays. There are three main structural components; a thin disc with a diameter of about 100 thousand light years and a thickness of about 3000 light years, a roughly spherical central bulge with a diameter of about 20 thousand light years, and an extensive halo of relatively low density and uncertain size. The whole structure is thought to be immersed in a cloud of 'dark matter'. The nature and distribution of this dark matter is not currently known, but its gravitational effects indicate that its total mass exceeds that of the visible matter by at least a factor of ten. There are about 100 thousand million stars in the Milky Way. The Sun is just one of those stars, probably located about 30 thousand light years from the Galactic centre, between two of the major spiral arms.

3.3 The Milky Way: a view from the Earth
Because we are located within the disc of the Milky Way – about a third of the way in from one edge – the stars and gas of the disc surround us on all sides. On a clear night, when the sky is dark, this encircling band of material can be seen arching across the sky. This particular image captures almost half of the Milky Way and was taken at a time when the Galactic centre was overhead. The Galactic centre itself cannot be seen at visible wavelengths because of the obscuring effects of the gas and dust in the plane of the disc.

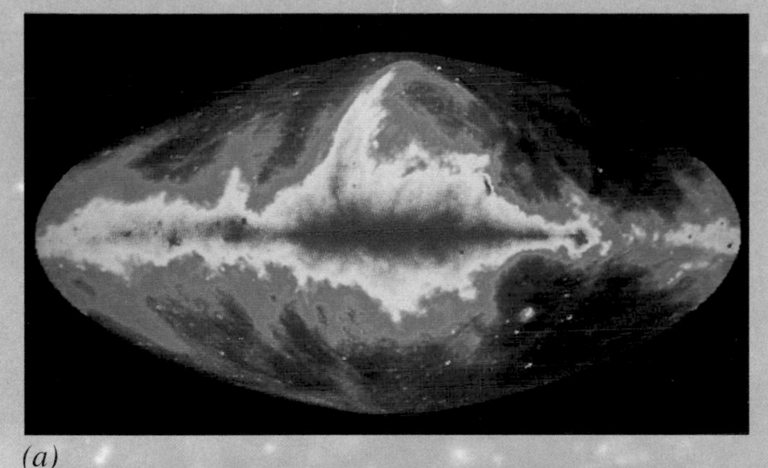

(a)

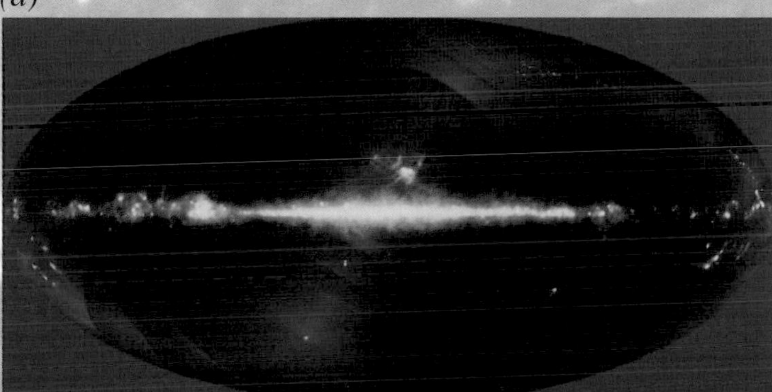

(b)

(c)

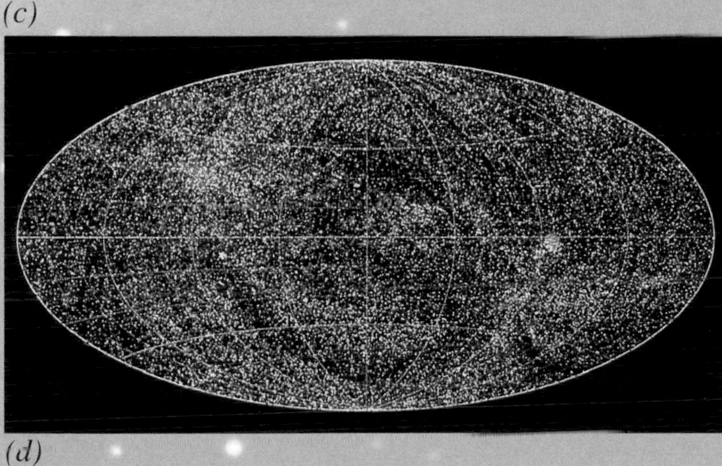

(d)

3.4 All-sky views of the Milky Way at many wavelengths
Each of these four images shows the whole sky (the entire celestial sphere), using a map projection that has the central plane of the Milky Way running horizontally across the middle of the picture. (a) At a wavelength of 73.5 centimetres the plane of the Milky Way is a relatively intense source of radio waves, as is the prominent northward pointing spur that may be a remnant from a nearby supernova. (b) At infrared wavelengths much of the emission is from cool dust, some of which is associated with star-forming regions, such as Rho Ophiuchi (just north of the Galactic centre) and Orion (south of the Galactic plane, at the extreme right). (c) At visible wavelengths, the obscuring effects of dust and gas limit the familiar view of the stars to a range of about 15 thousand light years in the plane of the Milky Way's disc, which has a total diameter of about 100 thousand light years. Away from the plane of the disc the view is not obscured and three nearby galaxies can be seen as fuzzy patches to the south.
(d) Preliminary data from the orbiting observatory Rosat are helping to clarify our view of the X-ray sky. X-ray sources within the Milky Way include supernova remnants, neutron stars and (perhaps) black holes, but there also many extra-galactic sources.

(a)

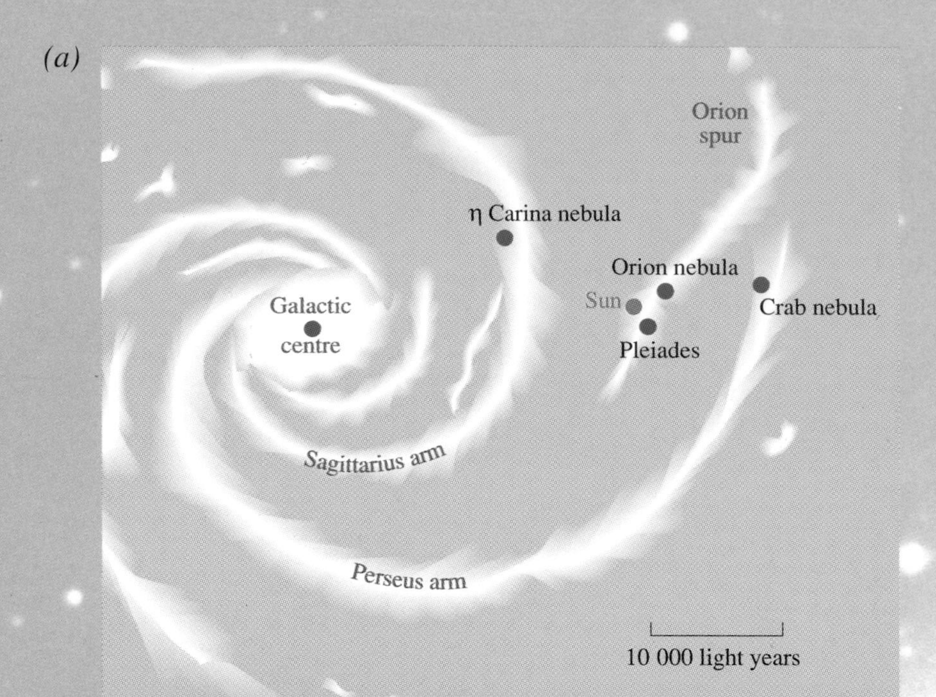
Orion
spur
η Carina nebula
Orion nebula
Sun
Crab nebula
Galactic
centre
Pleiades
Sagittarius arm
Perseus arm
10 000 light years

(b)

10 arcmin

(c)

10 arcmin

(d)

20 arcmin

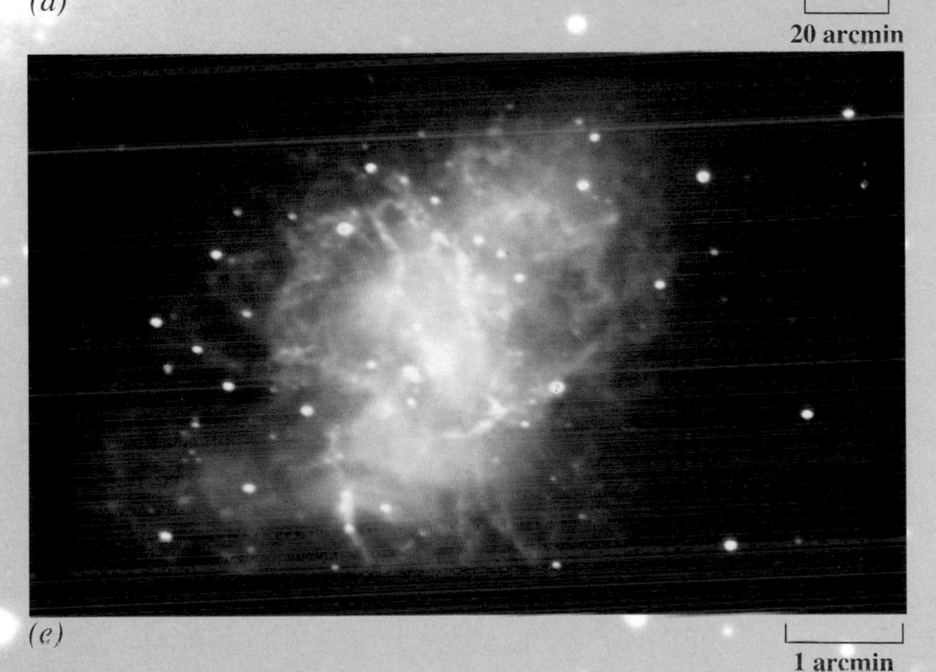

(e)

1 arcmin

3.5 Sights of the Milky Way's disc
(a) The discs of spiral galaxies are regions in which stars are continuing to form. In such discs, spiral arms stand out because they contain a higher than average proportion of bright young stars. Many signs of star formation can be seen in the disc of the Milky Way. (b) In the same spiral arm or spur as the Sun, at a distance of about 1630 light years, is the stellar nursery of Orion.
(c) Even closer, about 410 light years away, is the Pleiades open cluster – a group of several hundred young stars that probably emerged from their own stellar nursery less than 50 million years ago. (d) Closer to the centre of the Galaxy, roughly 9000 light years away, in the neighbouring Sagittarius arm, is the Carina Nebula – a convoluted cloud of gas illuminated by the bright young stars that it contains.
(e) In the outlying Perseus arm, at a distance of approximately 4000 light years, is the Crab Nebula – the remnants of a star that was seen to end its life in a spectacular supernova in the year 1054.

3.6 Sights of the Milky Way's halo
About 1% of the stars belonging to the halo are gathered together into 150 or so globular clusters. (a) The distribution of globular clusters is concentrated around the centre of the Milky Way. By determining the middle of this distribution the centre of the Galaxy can be located.
(b) M13, (c) M55 and (d) 47 Tucanae are three of the more prominent globular clusters. Typically, such clusters are about 150 light years in diameter and contain between a hundred thousand and a million stars. Long-exposure photographs that reveal the outer parts of globular clusters usually over-expose the centre. (e) However, an ultraviolet view, from the Hubble Space Telescope, reveals that even at the core of 47 Tucanae the stars are well separated. Globular clusters are generally very ancient, but some of the stars in this view are quite blue, indicating that they have been rejuvenated in some way, perhaps by interacting with binary companions.

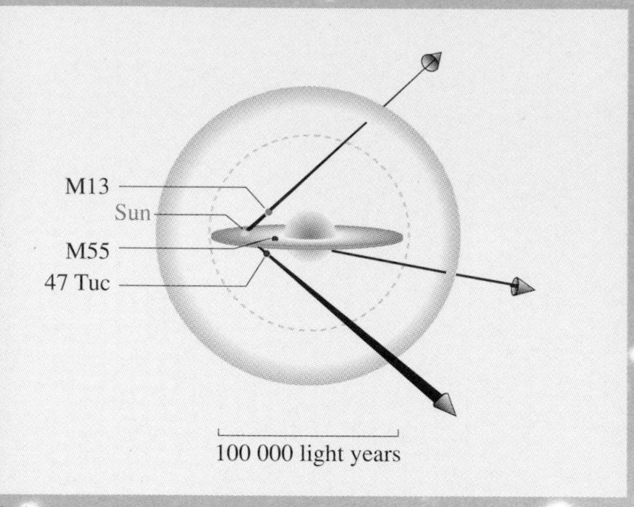

(a)

(b)

4 arcmin

(c)
10 arcmin
(d)
5 arcmin
(e)
2 arcsec

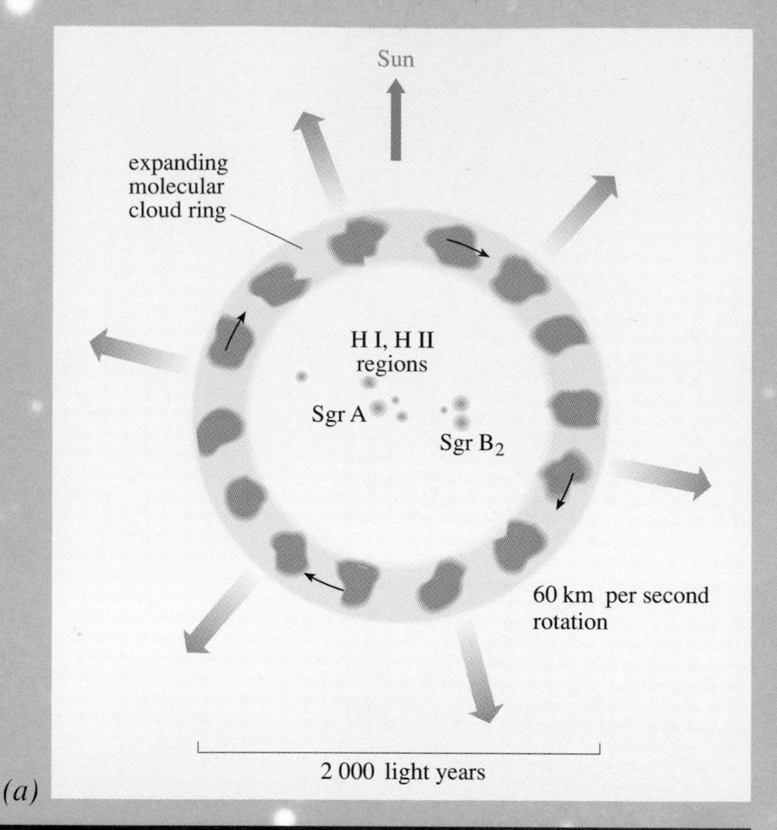

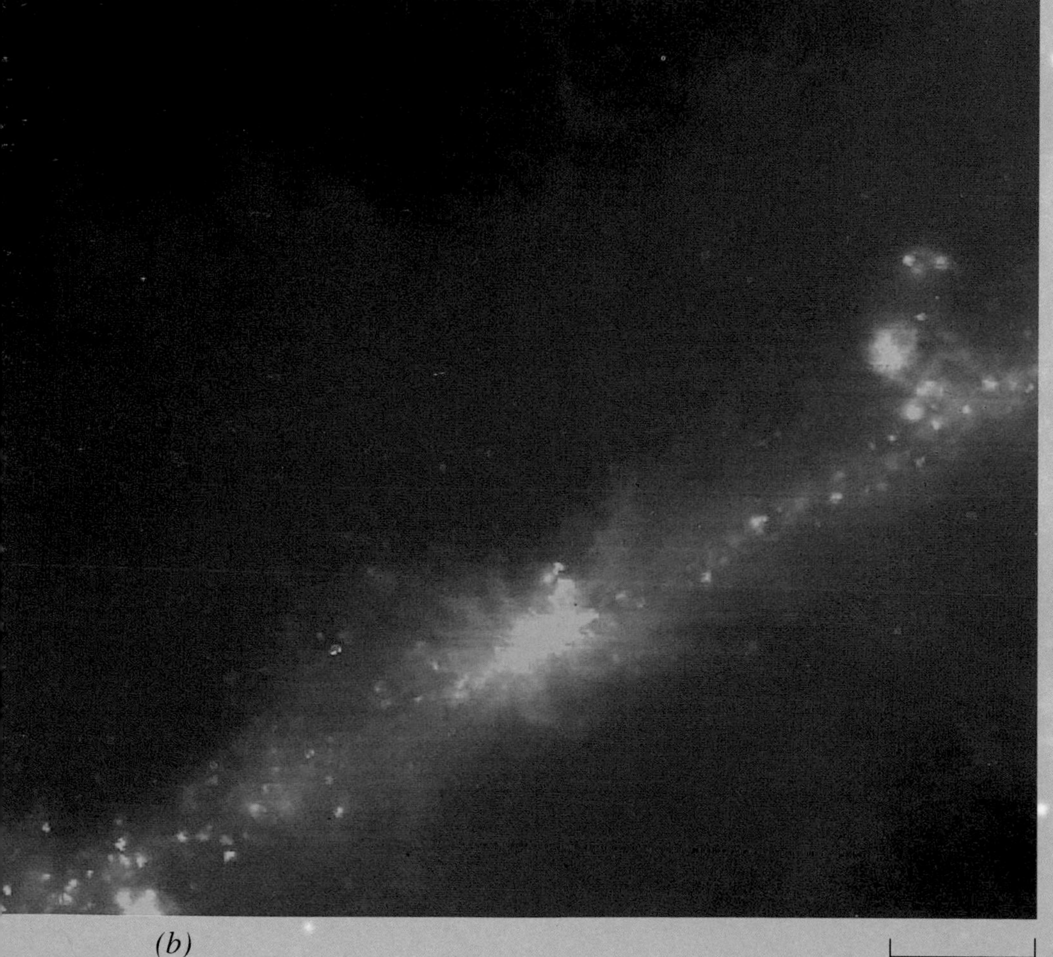

3.7 Sights of the Milky Way's bulge
(a) Light from the central regions of the Milky Way cannot be seen from the Earth owing to the obscuring effects of gas and dust. Nonetheless, studies at a variety of non-visual wavelengths have revealed the presence of complicated structures in the core of our galaxy.
(b) Infrared images such as this wide-angle view have played an important part in revealing the complexity of the Galactic centre. However, it is radio investigations that provide the most detailed views. (c) This image, based on observations at wavelengths of 6 and 20 centimetres, shows the 'mini spiral' that occupies the central 10 light years of the Milky Way. It has been suggested that the mini spiral may actually be a rotating ring of material surrounding two inflowing streams. Many investigators suspect that a super-massive black hole located at the centre of the Milky Way might be responsible for the mini spiral and for much of the other activity seen in the central bulge.

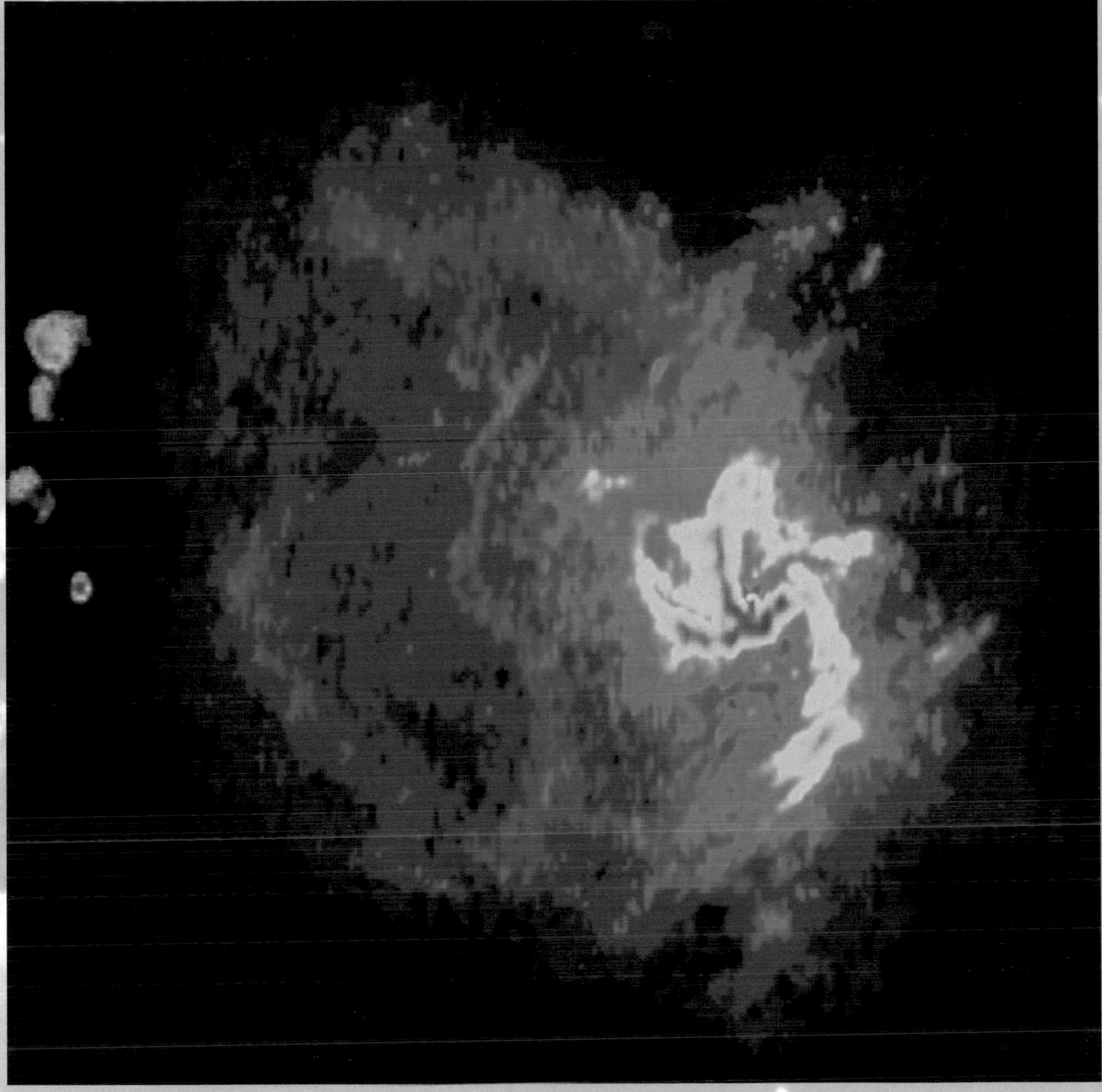

(c)

(a)

1 arcmin

3.8 Spiral galaxies
(a) The Sombrero Hat galaxy, M104 in Virgo, is a spiral galaxy of Hubble type Sa. It has a large central bulge and tightly wound spiral arms. Because we observe it almost 'edge-on' it is possible to see a dark lane that crosses its disc. This is caused by dust in the mid-plane. M104 is roughly 65 million light years away. It has a diameter of about 75 thousand light years, and takes about 200 million years to rotate. (b) M31 in Andromeda is an Sb galaxy and is probably similar to the Milky Way. It is a mere 2.4 million light years away and was the first object definitively shown to be an external galaxy, beyond the boundaries of the Milky Way.

(b)

20 arcmin

(c)

2 arcmin

(c) M83 in Hydra is an Sc galaxy at a distance of approximately 15 million light years. Inclined at about 60° to the line of sight, it has a relatively small bulge and wide-flung spiral arms. Bright regions of active star formation can be seen along the arms. (d) NGC 1365 is a barred spiral of Hubble type SBb. A part of the Fornax cluster of galaxies, its distance is about 55 million light years. Though visually striking, the presence of a central bar seems to have little influence on the overall properties of a spiral galaxy.

2 arcmin

(d)

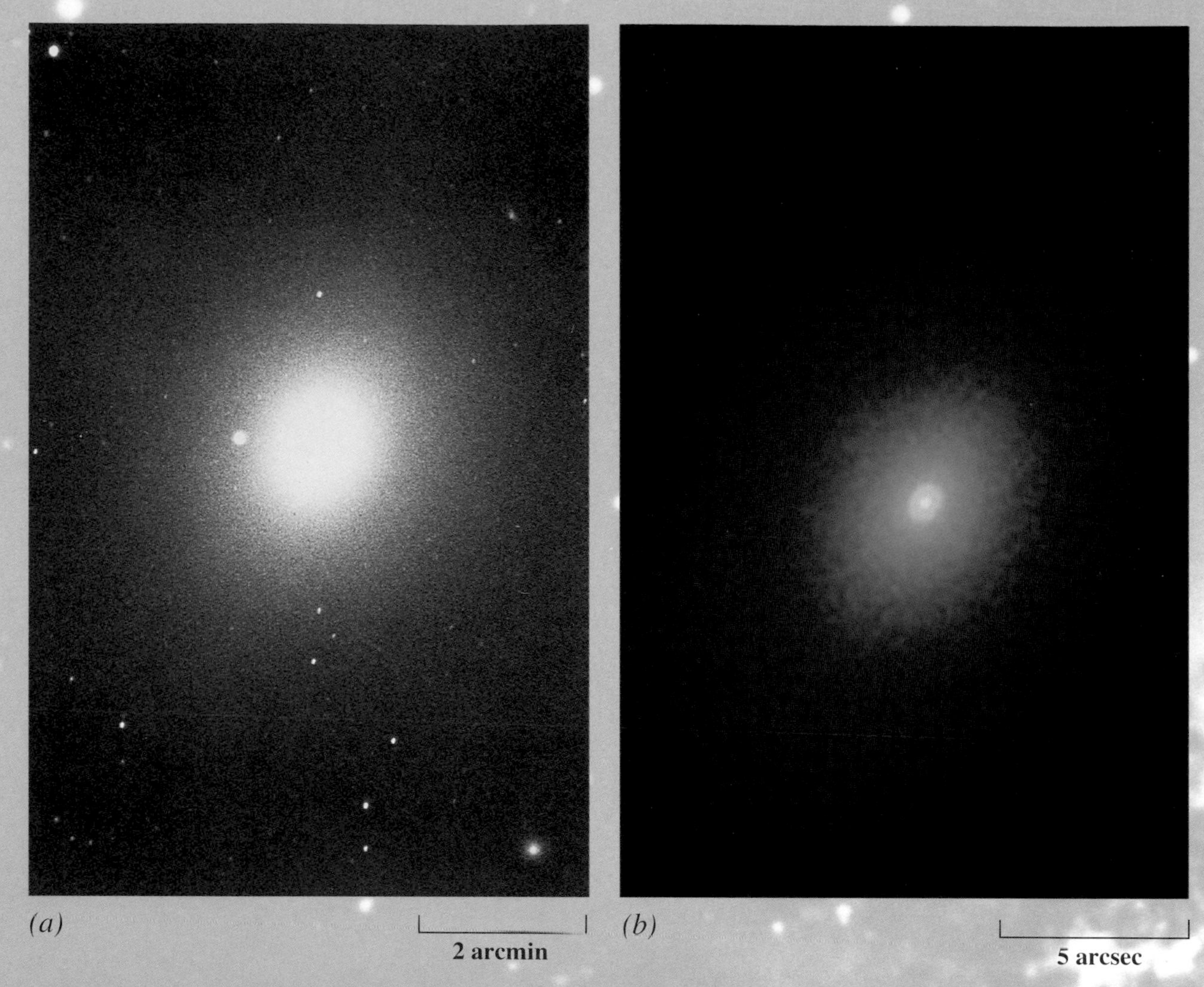

3.9 Elliptical galaxies
Elliptical galaxies come in a wider range of sizes and masses than spiral galaxies. Unlike spirals, elliptical galaxies contain little gas and show almost no sign of active star formation. (a) M49 in Virgo appears to be almost featureless and is classified as an elliptical galaxy. Its approximate distance is 55 million light years. (b) M32 is a small elliptical companion of the spiral galaxy M31. Somewhat elongated, it is classified as Hubble type E2. This Space Telescope view of the core of M32 shows that it, too, is elongated. (c) Leo 1 is a dwarf elliptical galaxy that is less than a thousand light years in diameter. At a distance of about 750 thousand light years, it is one of the ten nearest galaxies known. Such dwarf galaxies may be very common, but their small size and low luminosity makes them difficult to detect at any great distance.

(a)

2 arcmin

(b)

50 arcmin

(c)

50 arcmin

3.10 Irregular galaxies
(a) IC 5152 in Indus is an irregular galaxy. Such galaxies lack the grand design of spirals, but they contain substantial amounts of gas and are often active sites of star formation. Generally smaller than spirals, irregular galaxies cover a wide range of sizes and masses. At a distance of about 3 million light years, IC 5152 is a small galaxy with a correspondingly small mass. The bright star slightly to the north of the galaxy is part of our own Milky Way. (b) The Large Magellanic Cloud (LMC) is probably the Milky Way's nearest galactic neighbour. The distances of most galaxies are rather poorly determined but the LMC is an exception. Its distance is thought to be 168 thousand light years with an accuracy of about 5%. The LMC can be seen with the naked eye as a fuzzy patch about 6° across by observers in the southern hemisphere. (c) The Small Magellanic Cloud is another irregular companion of the Milky Way. Somewhat smaller and farther away than the LMC, it has a visible diameter of about 3.5°.

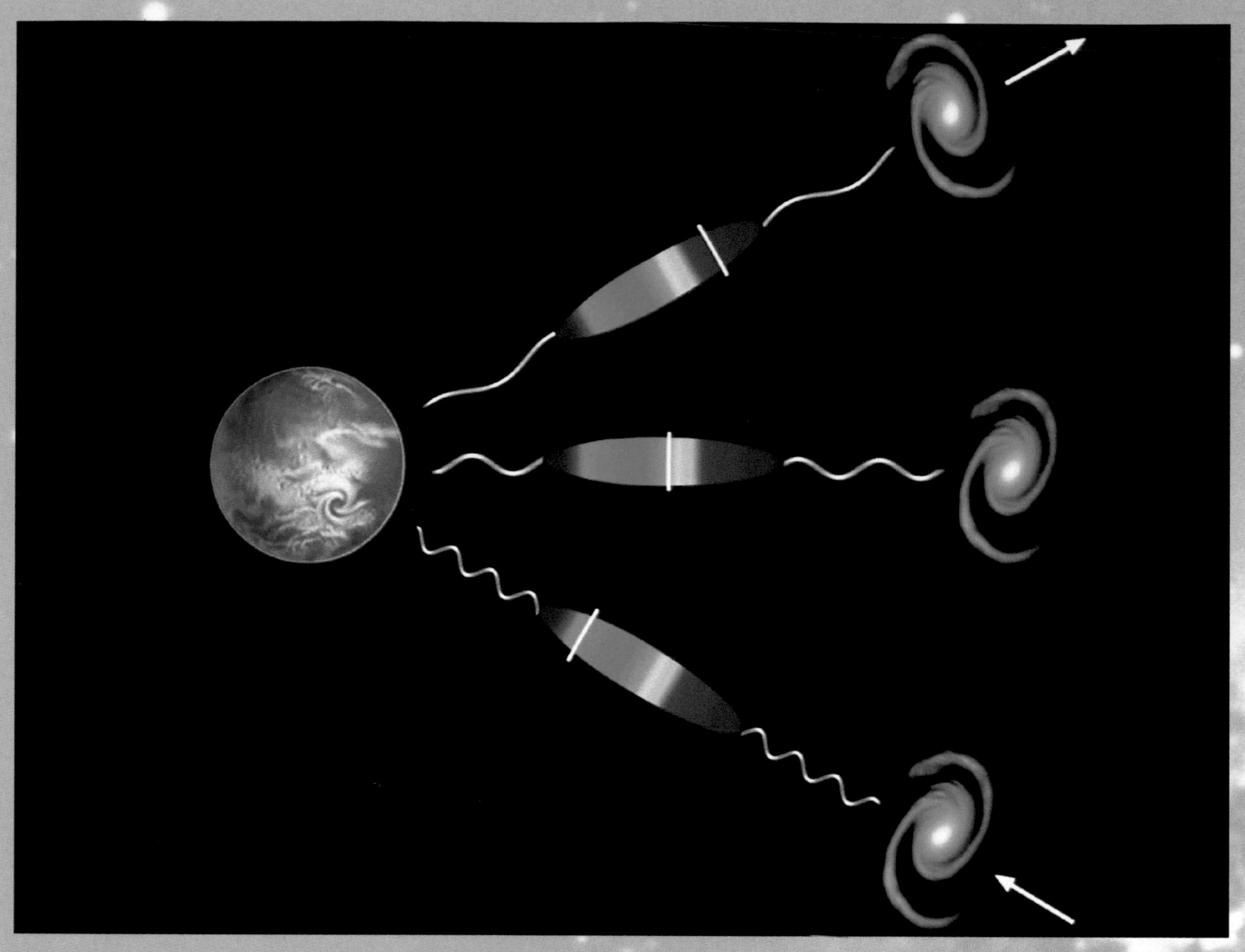

(a)

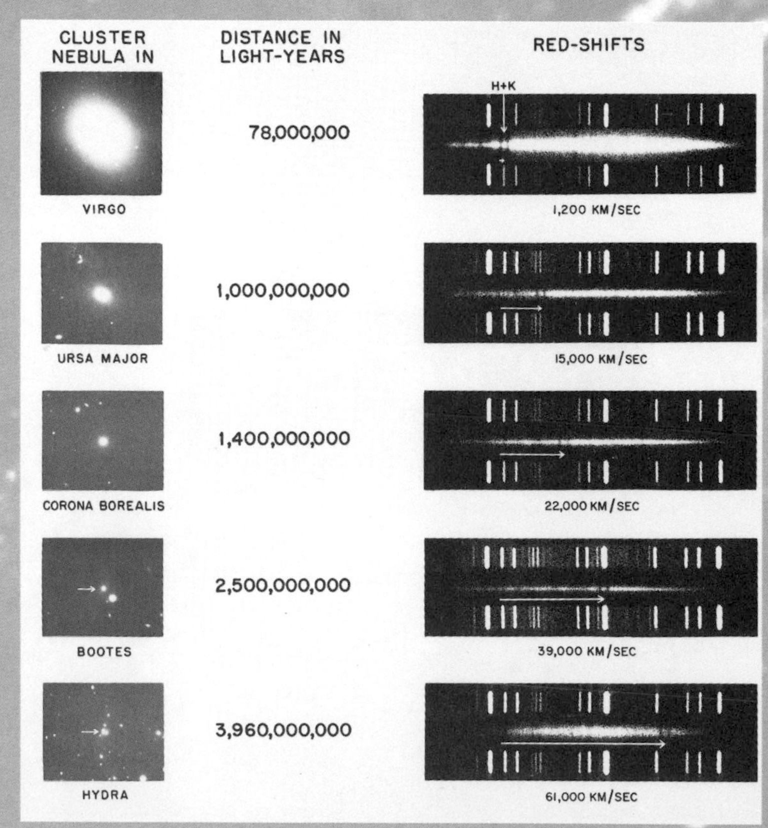
CLUSTER NEBULA IN
DISTANCE IN LIGHT-YEARS
RED-SHIFTS
H+K
VIRGO
78,000,000
1,200 KM/SEC
URSA MAJOR
1,000,000,000
15,000 KM/SEC
CORONA BOREALIS
1,400,000,000
22,000 KM/SEC
BOOTES
2,500,000,000
39,000 KM/SEC
HYDRA
3,960,000,000
61,000 KM/SEC

(b)

3.11 Redshift, Hubble's law and the distances of galaxies

(a) The wavelengths of spectral lines originating in galaxies can be stretched or compressed by effects such as the motion of the galaxy and the overall expansion of the Universe. Such changes cause spectral lines to shift towards either the red or the blue end of the spectrum. (b) The right-hand column of this famous image shows the spectra of the five increasingly distant galaxies shown in the left-hand column. The spectra are the cigar-shaped streaks; the lines above and below are for reference purposes. The H and K absorption lines due to ionized calcium can be seen in each spectrum, redshifted by an amount that increases roughly in proportion to the distance of the galaxy. Edwin Hubble interpreted these redshifts as a result of recessional motion and thus discovered the expansion of the Universe. Having determined the rate of cosmic expansion (Hubble's constant) it is possible to deduce the (approximate) distance of a far-off galaxy from its redshift. This technique is widely used, but uncertainties about the true rate of expansion mean that many galactic distances may be wrong by as much as a factor of two.

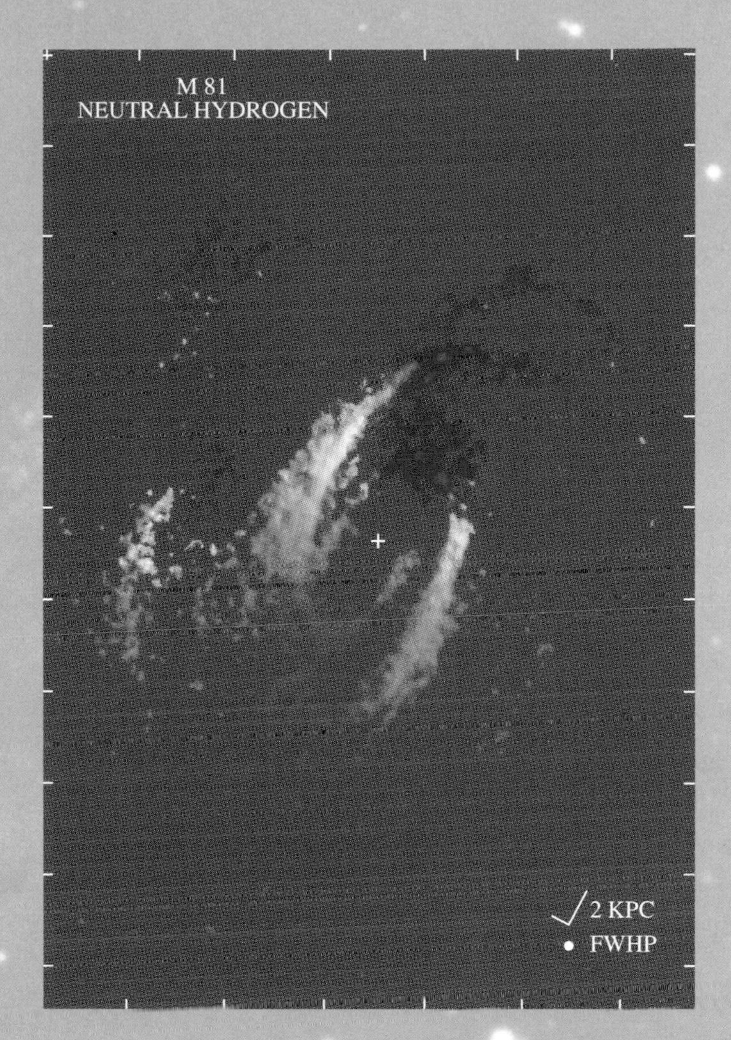

3.12 Redshift, rotation and the masses of galaxies

This is an artist's impression of the spiral galaxy M81 in Ursa Major, which is about 11 million light years away. The image, which has an angular width of about 35 arcmin, is based on observations of radio waves, emitted at a wavelength of 21 centimetres by neutral hydrogen atoms. It is a false-colour image: the brightness shows the intensity of the hydrogen emission, and the colour shows the extent to which the radio waves emitted from various parts of the galaxy's disc have been redshifted relative to the emissions from the centre of the galaxy. The variations in colour imply that the disc is rotating, with the blue regions approaching the Earth relative to the red regions, which recede. By studying the way in which a galaxy rotates it is possible to deduce the mass of that galaxy and the way in which the mass is distributed. Such studies indicate that many galaxies (perhaps all) contain substantial amounts of non-luminous 'dark' matter. Such matter probably extends far beyond the observable edges of the galaxies in which it is detected.

3.13 The Antenna Galaxy: a colliding pair

The galaxies NGC 4038 and NGC 4039 are colliding. Because stars are tiny compared with the spaces that separate them, it is quite possible that none of the individual stars in either galaxy will hit any other stars. Nonetheless, gravitational forces are causing large-scale distortions, and are almost certainly responsible for the two 'antennae' that give this object its name. Along with a number of other distorted galaxies, the Antenna Galaxy is listed in Halton Arp's catalogue of peculiar galaxies, hence it is also known as Arp 244. The Antenna Galaxy is about 83 million light years away. Its main body has a diameter of about 100 thousand light years.

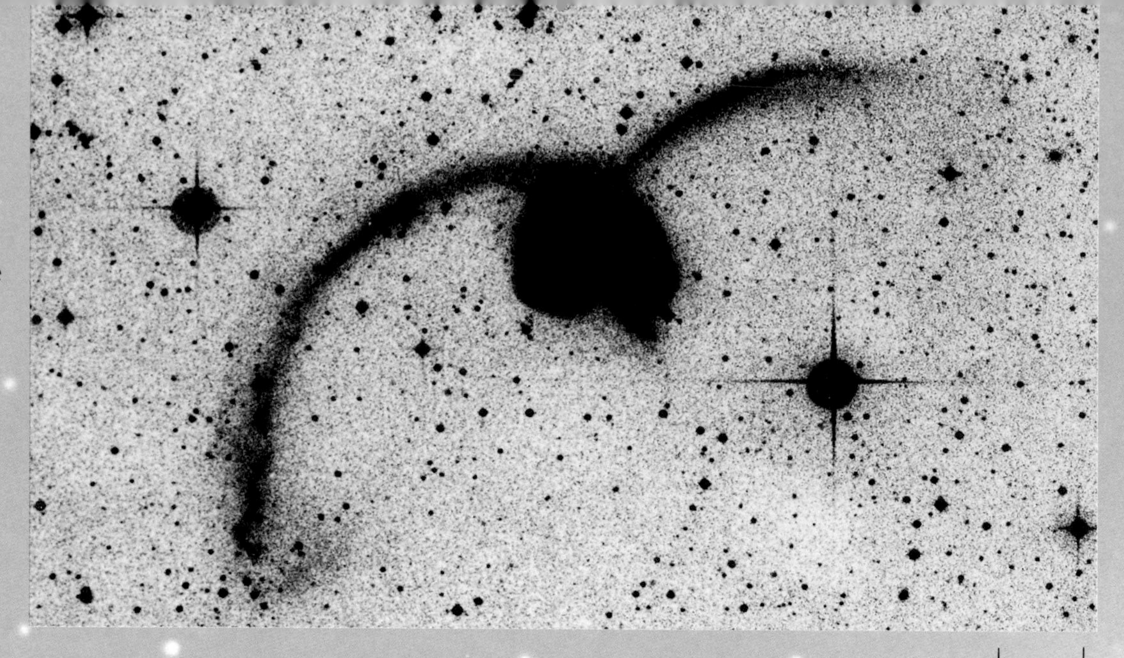

3.14 A computer simulation of colliding galaxies

In the early 1970s, Alar and Juri Toomre were able to simulate the collision of two galaxies using a computer. This sequence of 'snapshots' from their classic investigation shows that gravitational interactions alone can easily produce features such as the 'antennae' of NGC 4038-39 (Plate 3.13). Computer simulations are now widely used to study the complicated dynamic behaviour of colliding, merging and interacting galaxies.

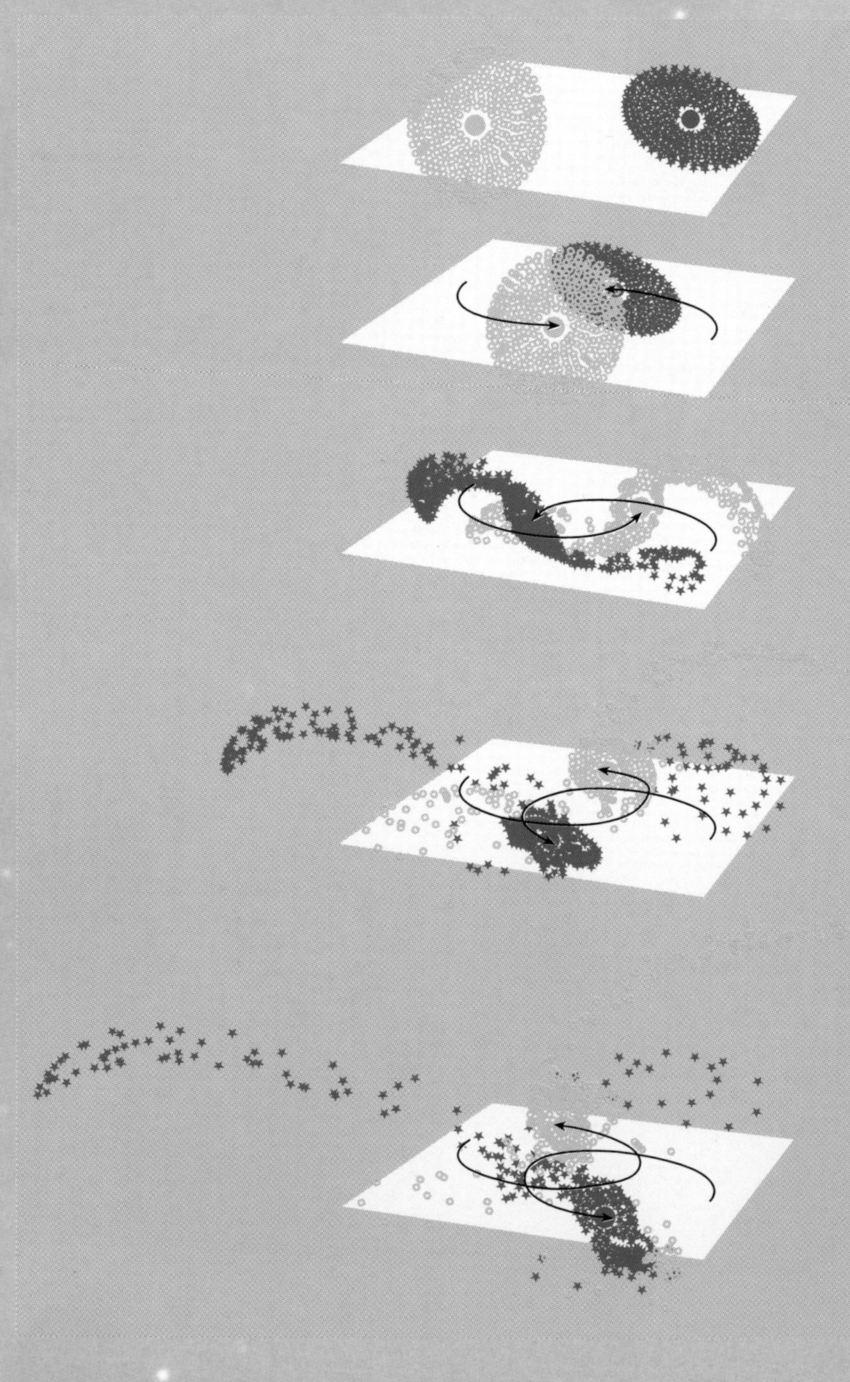

3.15 The Whirlpool Galaxy M51

M51, a peculiar Sbc galaxy located about 25 million light years away, is another example of a galaxy influenced by gravitational interactions. In this case, the source of the external gravitational force affecting M51 is the galaxy NGC 5195, visible just to the north of M51. The large angle of inclination of M51 gives us a good view of its spiral arms, which are unusually sharply defined. The arms, like those of other spiral galaxies, may be a consequence of spiral-shaped regions of enhanced density moving through the gas in M51's disc. As giant dark clouds, ready to participate in the process of star formation, enter such density enhancements they may become compressed and subsequently collapse and fragment. The star clusters emerging from such a collapse would include some highly luminous but short-lived blue-white stars. Such stars highlight the spiral arms. It may be that a companion galaxy or some other kind of gravitational disturbance is essential if a galactic disc is to develop spiral arms.

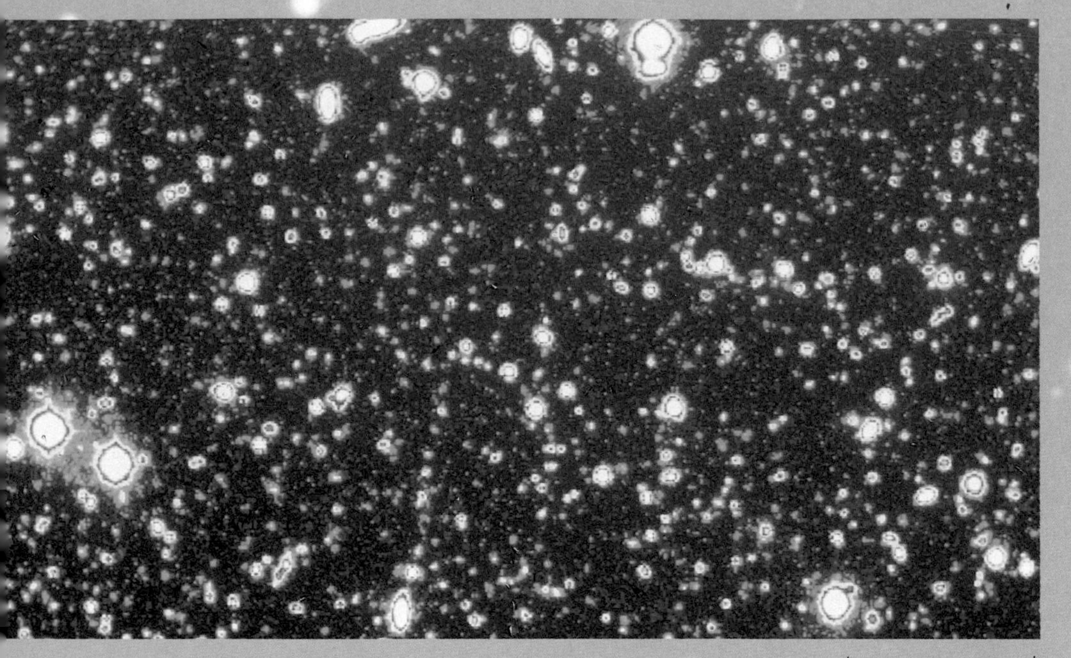

3.16 Galaxies at the limit of observation

This image of a relatively undistinguished patch of sky in the constellation of Pisces shows some of the faintest galaxies ever observed (28th magnitude). The picture is the result of 24 hours of observation at the Isaac Newton Telescope on the island of La Palma. It was recorded using a highly sensitive detector known as a CCD (charge coupled device). Studies carried out with such detectors, mounted on powerful telescopes, show that even apparently 'empty' regions of the sky are strewn with faint galaxies. When the faintest galaxies currently observable are taken into account, there are about half a million galaxies per square degree of sky.

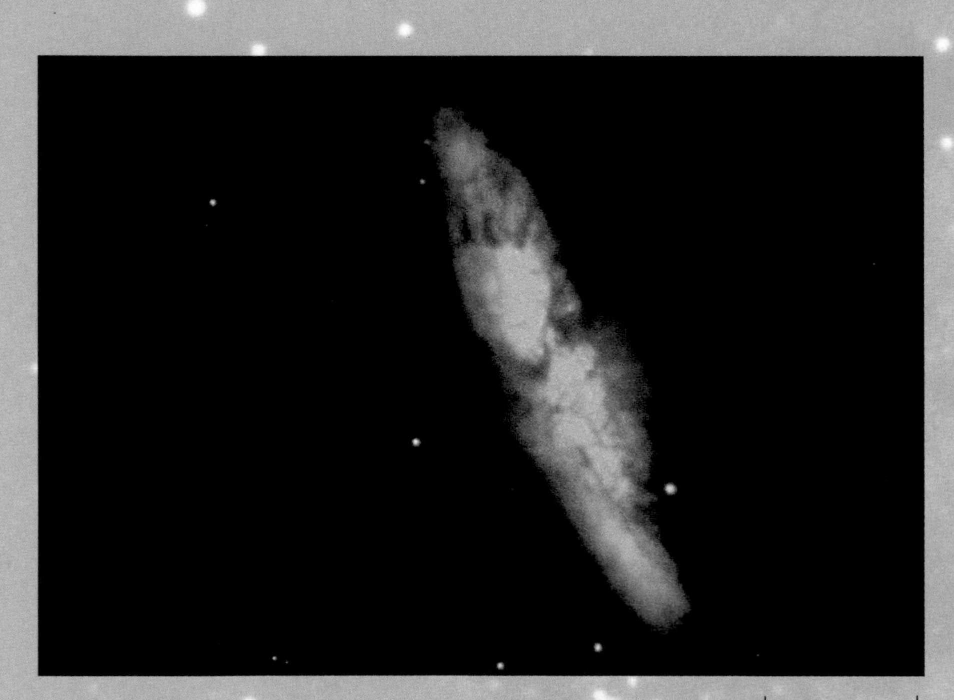

3.17 M82: a starburst galaxy

M82 has long been a puzzle to astronomers. At various times it has been described as an exploding galaxy, a colliding galaxy, or simply a peculiar irregular galaxy, but it is now assigned to the relatively new category of starburst galaxies. Like other members of the class, it is undergoing an enormous burst of star formation, hence the name. Starburst galaxies are powerful sources of infrared radiation because of their unusually high level of star formation. The origin of this activity is still under investigation, but is widely thought to be the result of an interaction with another galaxy. In the case of M82, which is at a distance of about 17 million light years, the cause may be a collision with a cloud of intergalactic gas produced by some kind of interaction with M82's nearby companion, the much larger M81 (Plate 3.12). Despite its ragged appearance, M82 itself seems to be a small spiral being viewed almost edge-on.

3.18 NGC 253: a starburst galaxy

Bursts of star formation can occur in any kind of galaxy, but NGC 253 is another spiral galaxy, like M82. A small Sc galaxy, at a distance of about 7 million light years, and inclined at just 17° to the line of sight, NGC 253 is somewhat smaller than the Milky Way. Like other starburst galaxies it is characterized by an extended region of infrared emission, and has a spectrum that shows strong emission lines. NGC 253 is the dominant member of a small group of galaxies known as the Sculptor Cluster. The Milky Way is also a member of a small cluster of galaxies, known as the Local Group, along with M31, the Magellanic Clouds and twenty or so other galaxies. The Sculptor Cluster is one of the nearest clusters of galaxies to our own Local Group.

3.19 Centaurus A: the nearest active galaxy

(a) Centaurus A is a powerful radio source located about 16 million light years away. Its outer emission lobes, mapped by radio astronomers, cover more than 9° of the sky, and have a total diameter of more than 2.5 million light years. Near the centre of this emission is a peculiar giant elliptical galaxy, NGC 5128 (Arp 153). Photographic studies reveal that this galaxy is surrounded by faint shells of material of the kind produced when galaxies merge together. Short photographic exposures (b) show a lane of dust crossing the central region of NGC 5128, at right angles to (c) the inner radio emission lobes. Although the total amount of radio energy emitted by Centaurus A is about a thousand times less than its light emission, it is still around a thousand times greater than the radio emissions from a normal spiral such as M31. Not all active galaxies are powerful radio sources (though many are) but they all have an unusually high energy output that cannot be explained simply in terms of the combined output of stars and gas.

(a)

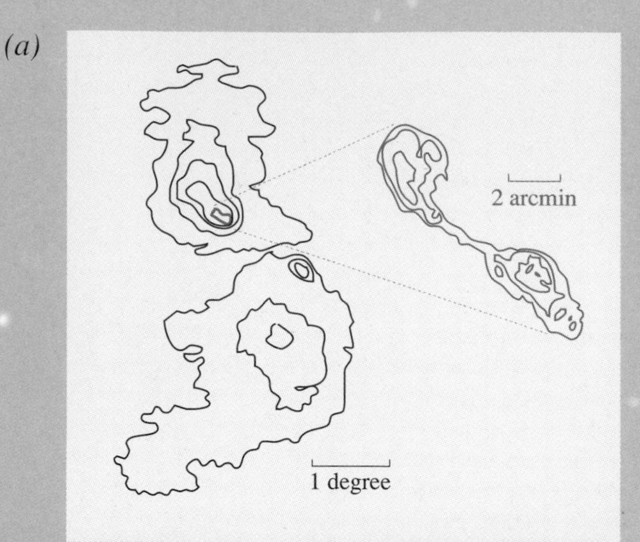

(b)

2 arcmin

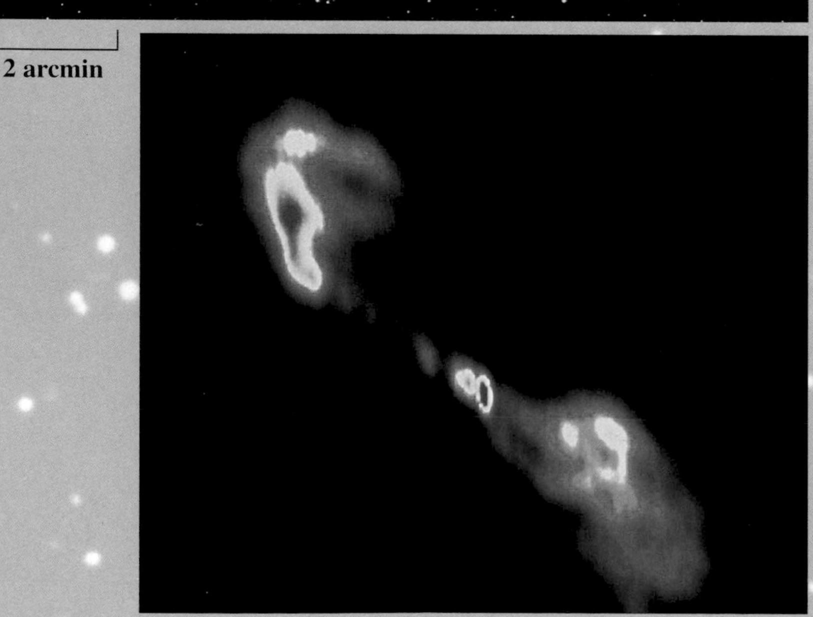

2 arcmin

(c)

3.20 Radio galaxies

(a) Cygnus A is another double-lobed radio source, rather like Centaurus A (Plate 3.19), though it is exceptionally powerful, outshining Centaurus A at radio wavelengths even though it is about 40 times further away. In the case of Cygnus A it is easy to see that the radio lobes are produced by beams of high speed particles jetting away from the peculiar optical source (b) at its centre. Broadly speaking, Cygnus A is a larger and more massive version of Centaurus A and is probably powered in the same way, perhaps by a super-massive black hole at the centre of the optical source.
(c) NGC 1265 is another radio galaxy, of a kind known as a narrow angle tail source. It is thought that the beams of charged particles ejected from its centre are swept back by its 2000 kilometres per second motion through the intergalactic gas of the Perseus Cluster. The Perseus Cluster is about 330 million light years away, far beyond the Local Supercluster of galaxies that contains the Local Group and the Sculptor Cluster.

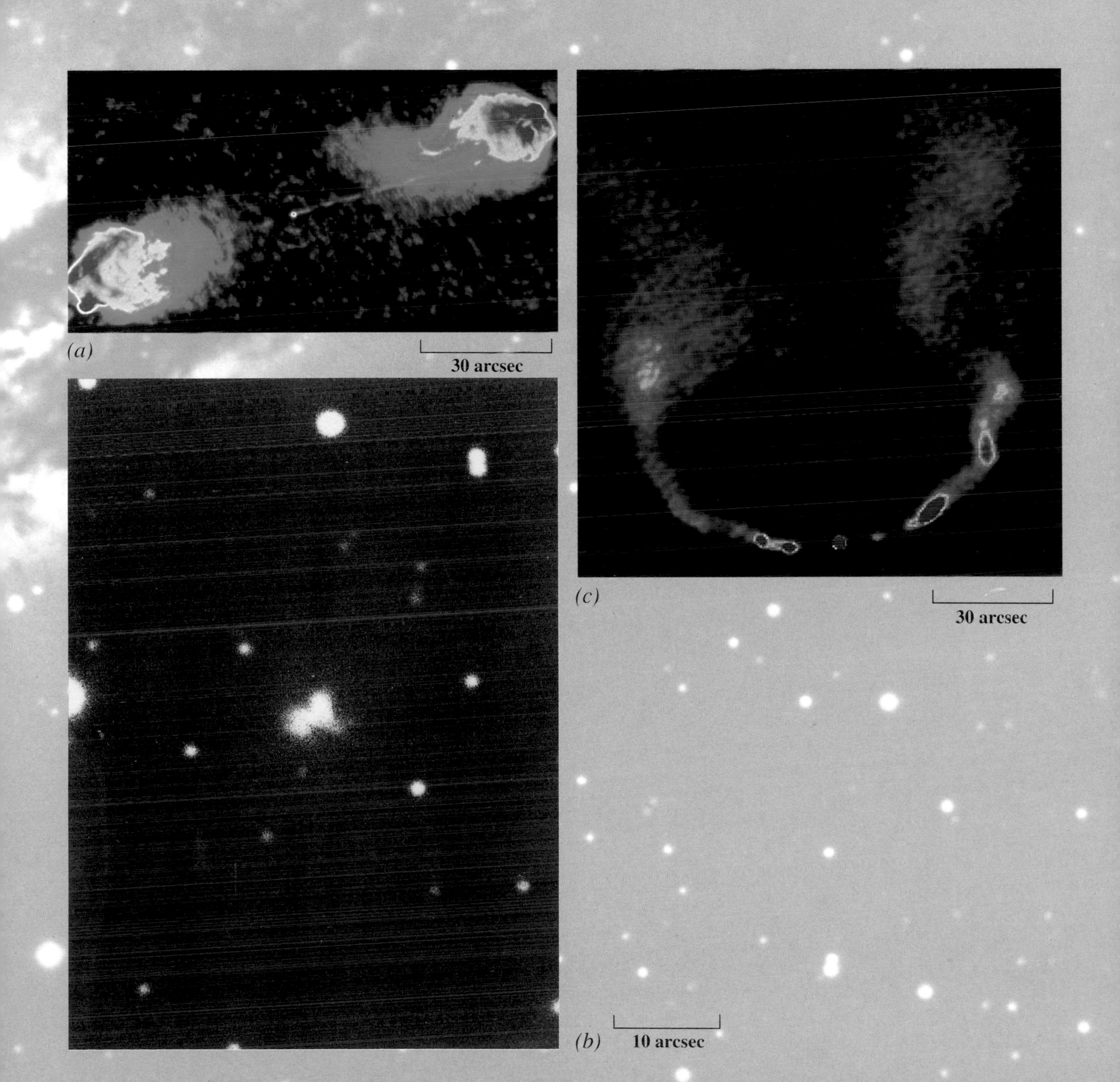

(a)

(c)

(b)

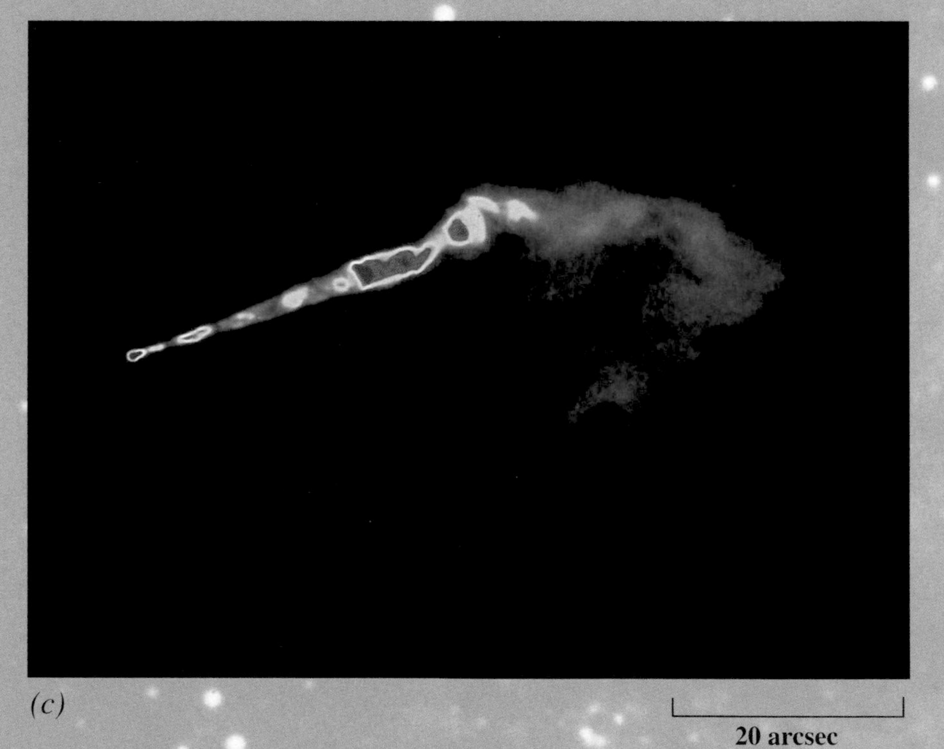

3.21 M87: a giant elliptical galaxy

About 55 million light years away, in a direction almost at right angles to the plane of the Milky Way, is the giant elliptical galaxy M87 (NGC 4486). Classified as a peculiar E0 galaxy, it is located in the dense Virgo Cluster of galaxies, close to the centre of the Local Supercluster. (a) Ordinary long-exposure optical images of M87 indicate a very large diameter and show hundreds of globular clusters. More specialized observations demonstrate that its diameter actually exceeds a million light years and imply that its mass is something like 30 million million times the mass of the Sun. (b) This near-infrared image of the central region of M87, recorded by the Hubble Space Telescope, reveals the galaxy's famous jet and the strong concentration of stars at its centre. (c) The fact that M87 is a radio source (Virgo A), combined with its other unusual features, gives weight to the suggestion that it may contain a black hole with a mass several million times that of the Sun.

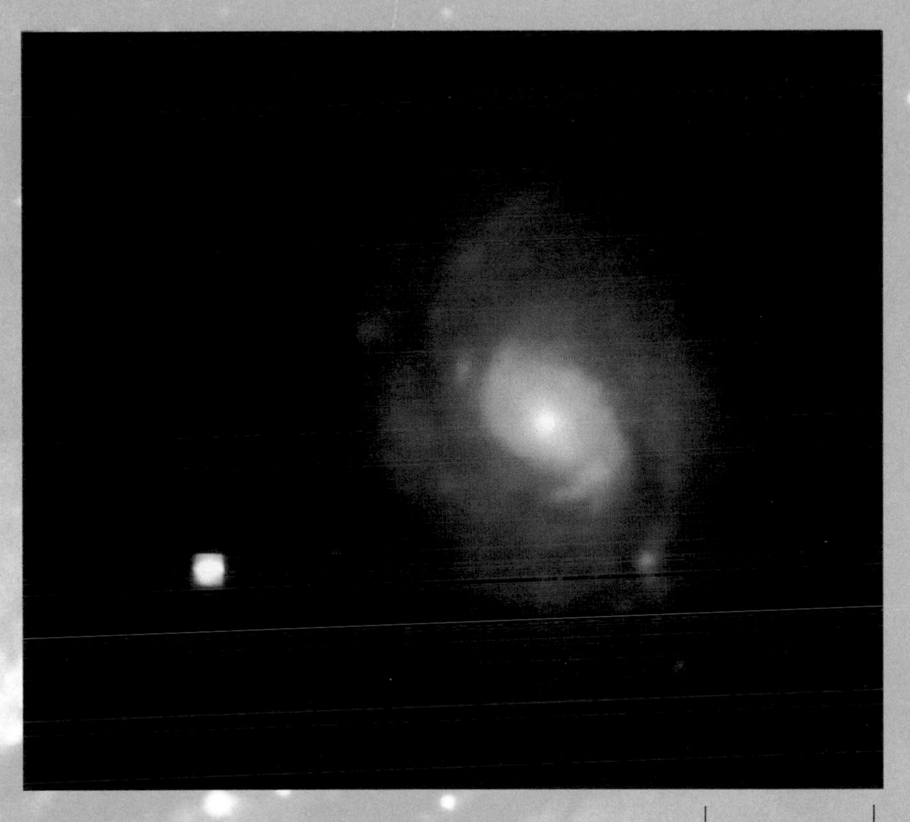

2 arcmin

3.22 *M77: a Seyfert galaxy*

M77 (NGC 1068) is a spiral galaxy of Hubble type Sb, about 47 million light years from the Milky Way. It is inclined at about 40° to the line of sight and its mass has been estimated at 27 thousand million times the mass of the Sun. M77 is a member of a class of active galaxies known as Seyferts, after their discoverer Carl Seyfert. In common with other members of the class, it has an abnormally bright core that fluctuates in brightness. Like many other Seyferts, M77 is a radio source, though not a particularly powerful one. It is quite clear that M77 contains an active galactic nucleus, a property it shares in common with other kinds of active galaxy, such as quasars (Plate 3.23). Indeed, it has been suggested that if they were farther away, some of the more powerful Seyferts would be almost indistinguishable from certain types of quasar. About 2% of spirals are Seyferts, making them the most common kind of active galaxy.

5 arcmin

3.23 *3C 273: the nearest quasar*

At visible wavelengths most quasars resemble faint variable stars and are distinctly unimpressive – hence the name quasar, a contraction of quasi-stellar object. However, their substantial redshifts imply that they are very far away and must therefore be very powerful to be seen at all. At a distance of about 2000 million light years, 3C 273 is one of the closest quasars – perhaps *the* closest. Its designation (3C 273) shows that it is a radio source, listed in the third Cambridge catalogue of radio sources. But, as this image from the orbiting Einstein X-ray observatory shows, it is also an X-ray source. The source to the north and east of 3C 273 is another quasar, even farther away, discovered by the Einstein observatory. Its redshift indicates that its distance is about 10 thousand million light years. Quasars are amongst the most distant objects than can currently be observed.

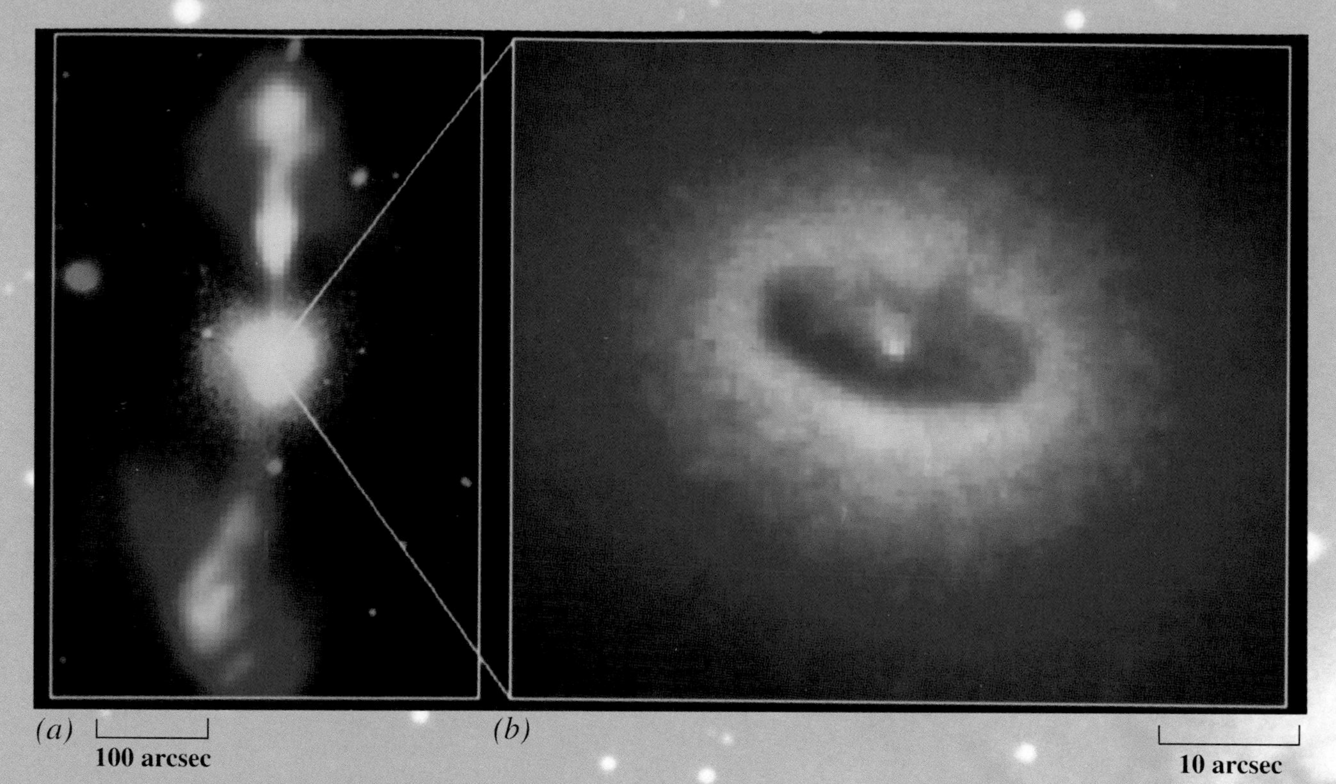

3.24 NGC 4261: a galactic core with a black hole?

NGC 4261, a peculiar E2 galaxy in the constellation of Virgo, is thought to be about 50 million light years away. (a) A combination of ground-based studies at visible and radio wavelengths shows opposed radio jets, spanning a total distance of about 150 thousand light years, emanating from the core of the galaxy. (b) A closeup of the core, obtained at visible wavelengths by the Hubble Space Telescope, shows a disc of gas and dust, inclined at an angle of about 30°. It is suspected that the observed disc is rotating around a massive black hole which it is gradually feeding with matter. Such black holes might be a common feature of active galaxies.

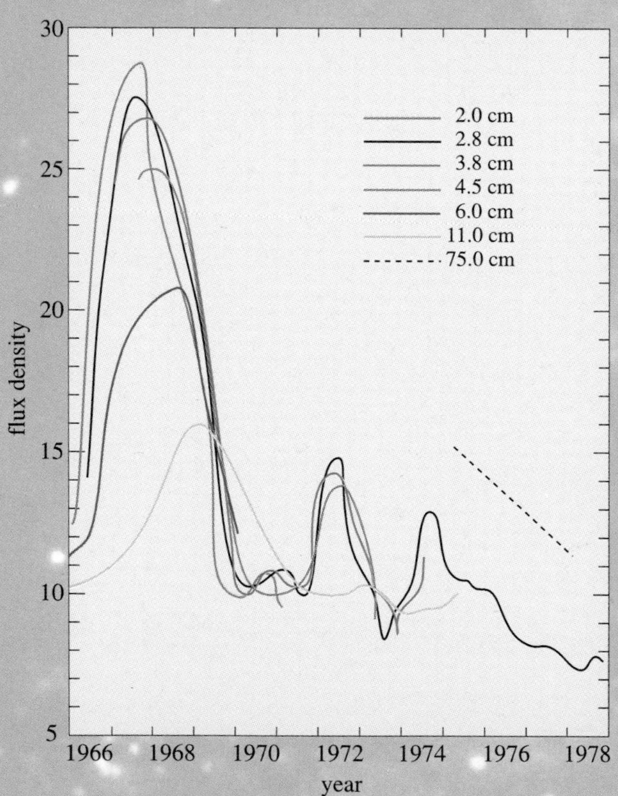

3.25 Variations in the radio energy from 3C 454.3

A characteristic of active galaxies is that their energy output varies with time. This graph, showing variations in the output from the quasar 3C 454.3 at various radio wavelengths, indicates that the quasar is the site of violent activity, Moreover, the period of time over which the output shows significant variations (months in this case) makes it possible to estimate the maximum possible size of the region responsible for the outburst. To account for coherent changes in output over a particular time t, the diameter of the source region cannot be much larger than the distance light can travel in that time. Thus, the 'engine' driving 3C 454.3 can be no more than about a light year across, and may be much smaller.

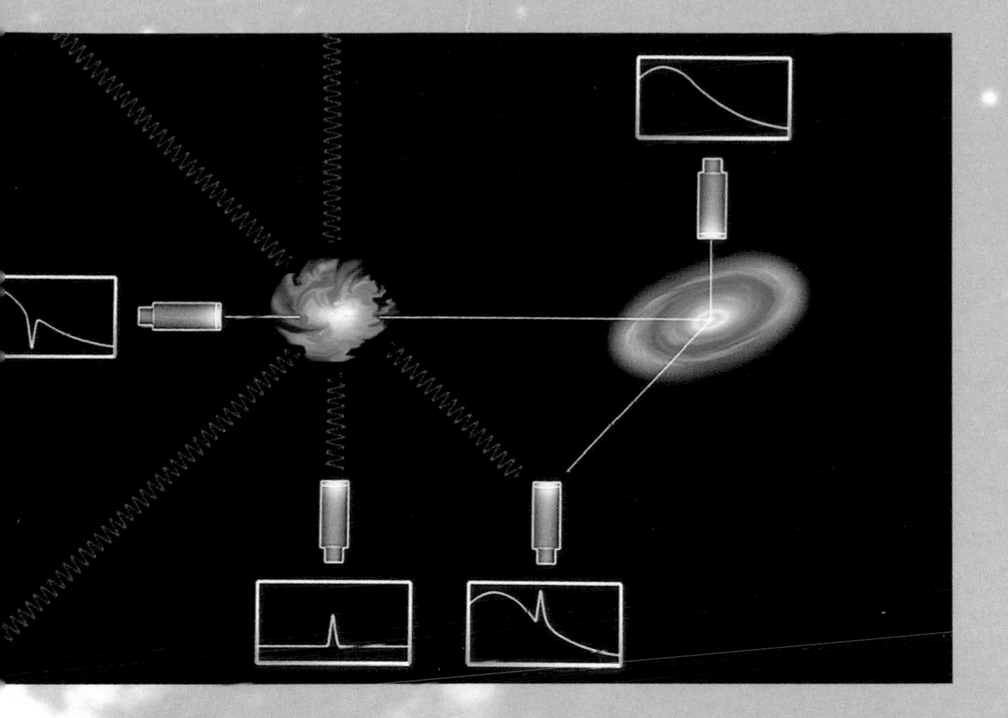

3.26 Origin of spectral features in an active galaxy

Whatever the nature of the 'central engine' of an active galactic nucleus, the various kinds of electromagnetic radiation that it emits must pass through the surrounding parts of the host galaxy before they can escape into the emptiness of space. During this process the spectrum of the emitted radiation can be modified in a number of ways. For instance, if the central engine is viewed through a gas cloud (shown in blue) the spectrum may show an absorption line at some characteristic frequency, and the cloud itself, viewed from an appropriate direction, can contribute emission lines to the spectrum (see Plate 1.4). In general, active galaxies might be expected to show either or both of these features in their spectra.

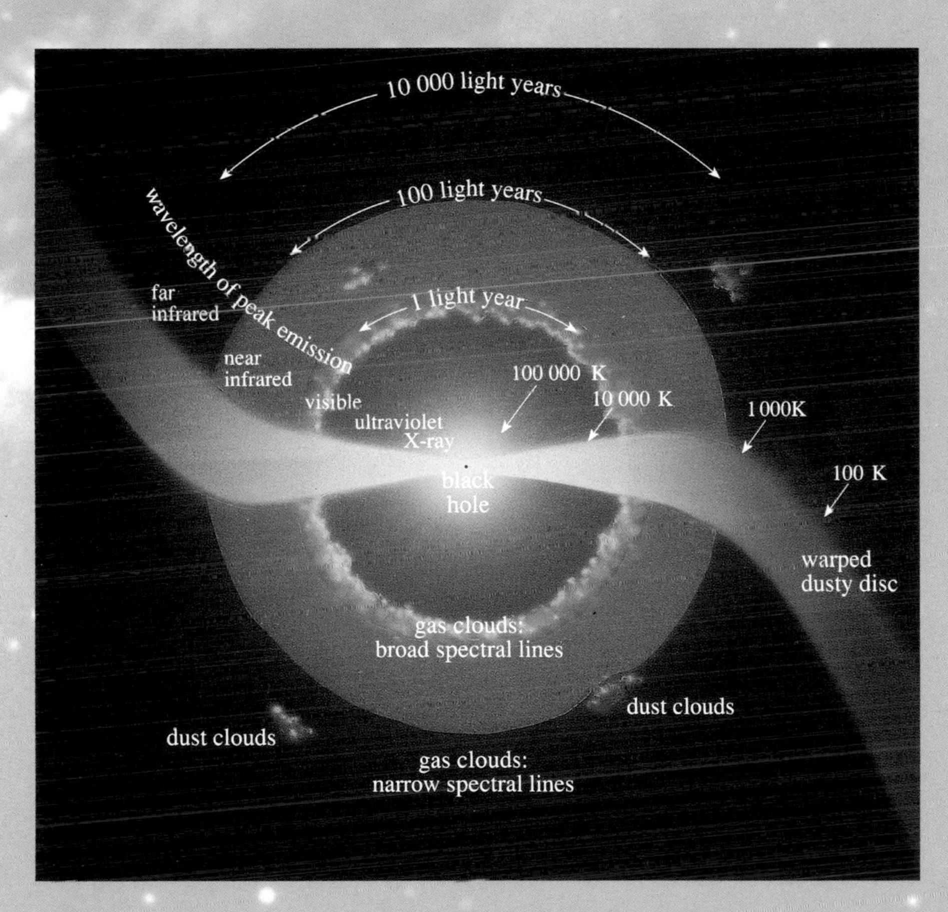

3.27 A model of a quasar

Many quasar features, such as enormously high luminosity and the presence of narrow and broad emission lines, can be explained by this theoretical model. The image needs careful interpretation because its scale varies with distance from the centre, each of the arcs at the top representing an increase in radial distance by a factor of 100. At the centre of the quasar is a massive black hole, acquiring hot matter from a surrounding disc. It is this hot accretion disc, probably no more than a light year across, powered by the black hole, that is the source of the X-rays and much of the other radiation that comes so copiously from the quasar. Surrounding the accretion disc, at a distance of a few light years, are moving gas clouds that contribute the spectral lines – at least those that originate within the quasar. Away from the centre, the accretion disc blends into a warped disc of warm dusty matter, which is shown in cross-section. The outer disc may extend for several thousand light years before it merges with the interstellar medium of the host galaxy that contains this active nucleus.

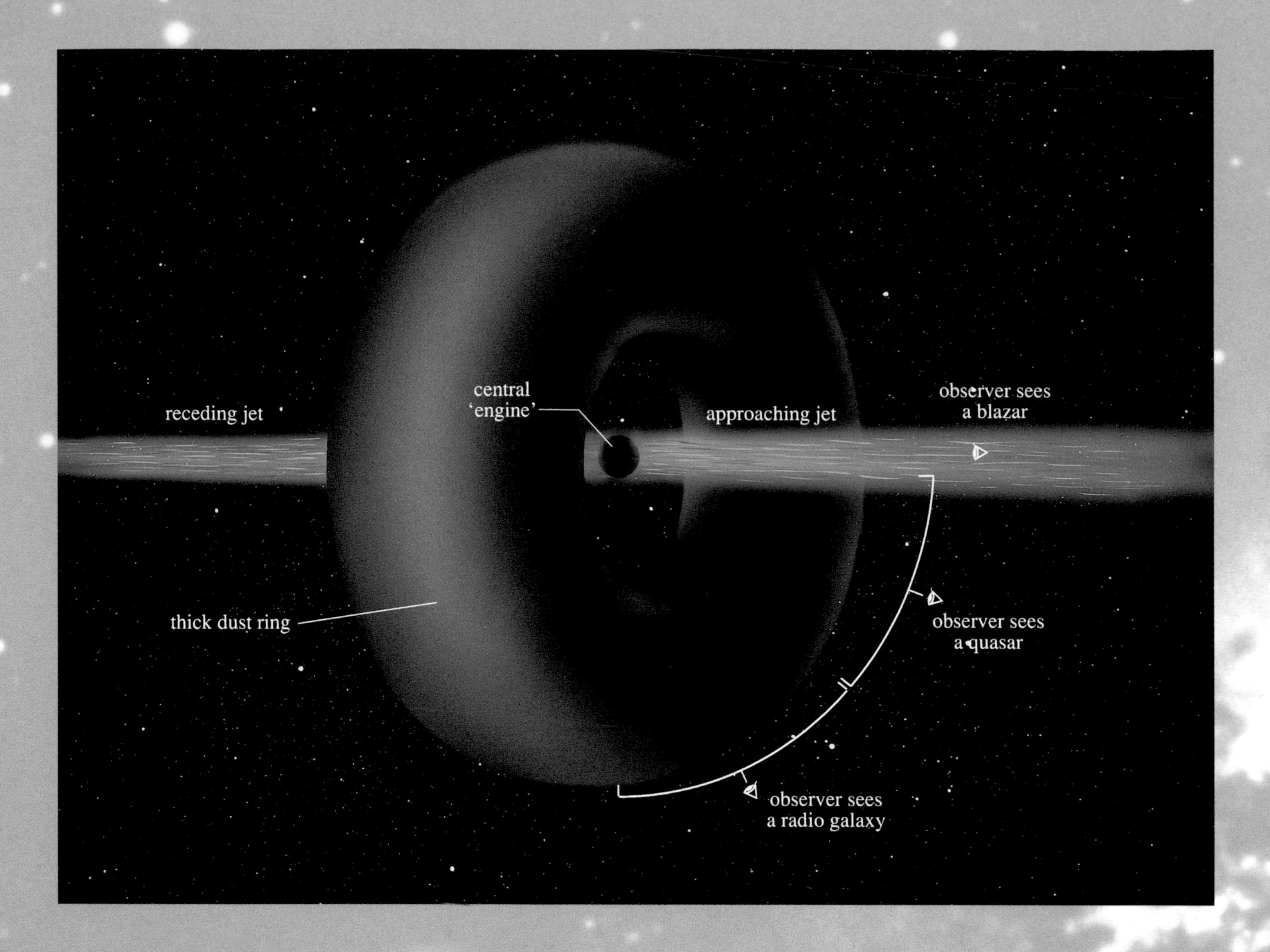

3.28 Unified model of active galaxies

There are many different kinds of active galaxy; blazars (BL Lacertae objects and optically violent variable galaxies), quasars, radio galaxies and so on. It has often been suggested that all these types might be fundamentally the same sort of object, simply viewed from different directions and shrouded by different amounts of gas and dust. This view is illustrated in this 'unified model'. The jets, which would be responsible for the twin lobes seen in radio galaxies, are expected to emerge at right angles to the plane of the central accretion disc, along the path of least resistance. Despite intensive research, the exact cause of such jets remains a mystery, though magnetic fields are widely thought to play an important part in their production.

3.29 The Oxford survey of galaxies
This image, based on an analysis of photographic plates from the UK Schmidt Telescope, covers about 10% of the sky (4300 square degrees) around the region of the southern galactic pole, in the constellation of Sculptor. The image is composed of 8 arcmin by 8 arcmin pixels, which have been shaded according to the number of galaxies they contain above a certain level of brightness. White pixels correspond to regions containing 20 or more galaxies, black corresponds to no galaxies and grey to numbers between 1 and 19. (Some regions have necessarily been omitted from the survey.) The resulting picture indicates that, on a size-scale of a hundred million light years, the distribution of galaxies is very uneven. Gigantic superclusters can be seen as large bright patches, often elongated, containing many smaller clusters, and surrounded by relatively dark regions that correspond to enormous voids. Despite the existence of superclusters with diameters of 100 million light years, on much larger scales the distribution of galaxies seems to be uniform with one patch of sky more or less indistinguishable from any other.

3.30 A part of the Virgo Cluster of galaxies
In almost the opposite direction to the Oxford Survey, not far from the northern galactic pole, is the rich cluster of galaxies in Virgo. The Virgo Cluster contains about a thousand galaxies, including the giant elliptical M87 (Plate 3.21), and occupies a region of space about 10 million light years across. At a distance somewhere between 40 and 80 million light years, the Virgo Cluster is the centre of the Local Supercluster of galaxies. Our own galaxy is a member of a sparse cluster known as the Local Group of galaxies. With 25 or so members and a diameter of about 7 or 8 million light years, the Local Group is an outlying member of the Local Supercluster. This image shows a region of the Virgo Cluster that is northwest of the cluster's centre. It includes a number of bright galaxies, such as the S0 galaxy M86 (near the centre) and the E1 galaxy M84 (at the right), but M87 is outside the field of view.

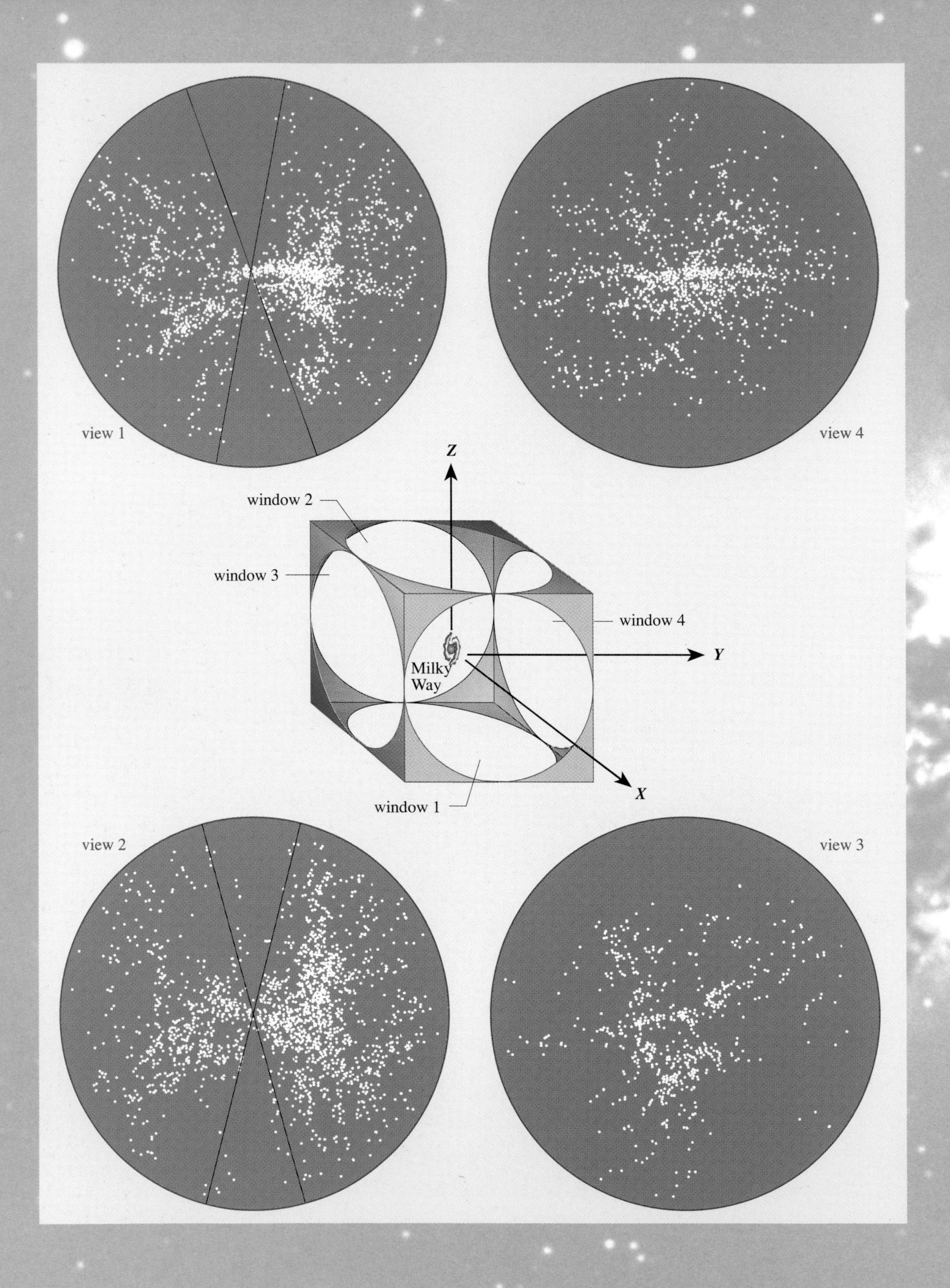
view 1
view 4
Z
window 2
window 3
window 4
Y
Milky Way
X
window 1
view 2
view 3

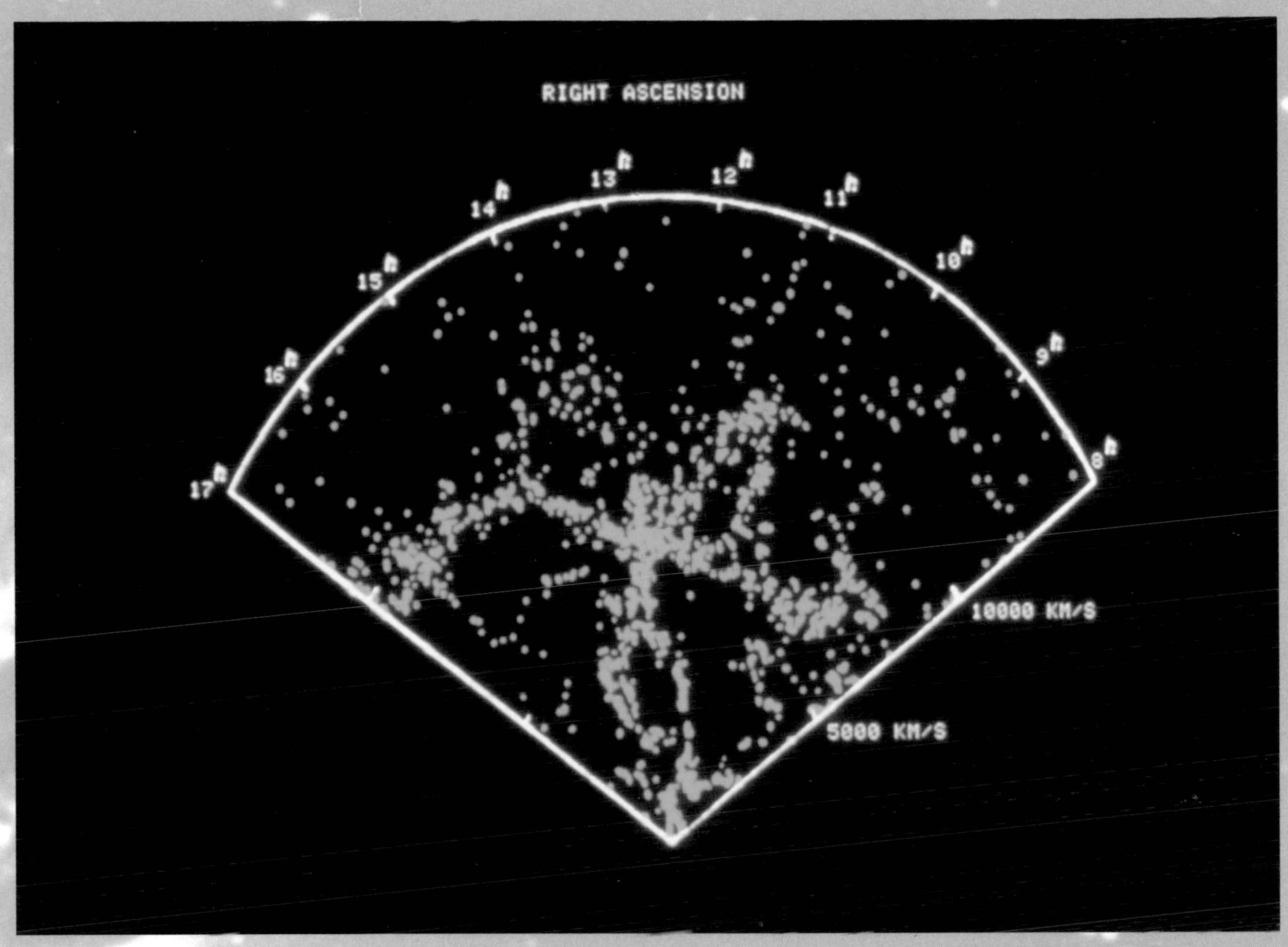

3.31 The distribution of 'nearby' galaxies

The three-dimensional distribution of relatively bright galaxies within about 100 million light years can be visualized by imagining them to occupy a 'fishtank', and then viewing the fishtank from a number of different locations. The Milky Way is located at the centre of the tank with its plane almost in the *XZ* plane and its north pole pointing in approximately the *Y* direction. Views 1 and 2 are views into the tank from outside and have the Milky Way at the middle. (The obscuring effects of the Milky Way's disc prevent observations close to the *XZ* plane, as shown by the empty triangular regions.) Views 3 and 4 are from the Milky Way looking south and north, respectively. The dense Virgo Cluster of galaxies, the centre of the Local Supercluster, can be clearly seen in views 1, 2 and 4.

3.32 The Stick Man

By measuring substantial numbers of redshifts and using Hubble's law to determine distances, it is possible to extend knowledge of the three-dimensional distribution of galaxies beyond the Local Supercluster. An extensive survey by members of staff at the Harvard-Smithsonion Centre for Astrophysics (CfA) has determined the distribution of galaxies in a number of 'slices' of the Universe. This image shows the result of examining one such slice, a narrow rectangular region of sky, 135° by 6°. The Milky Way is located at the point of the wedge. Other galaxies make up a feature known as the Stick Man that can be seen stretching away from us to encompass part of the neighbouring Coma Supercluster at a distance of about 300 million light years.

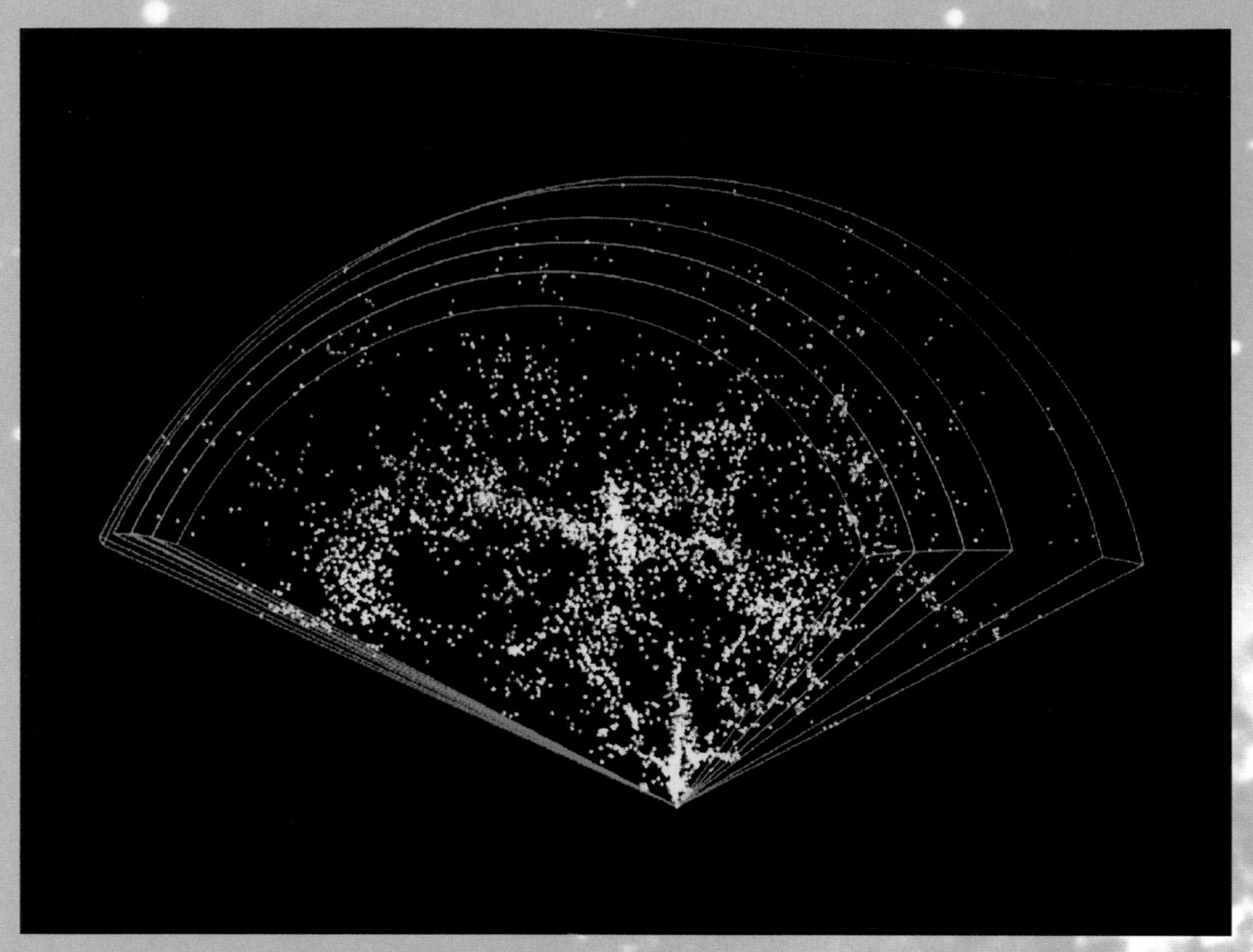

3.33 The Great Wall

By putting many 'slices' together, the CfA survey that revealed the Stick Man can also provide insight into even larger structures. Once again, the Milky Way is at the point of this wedge with part of the Local Supercluster nearby. However, the most prominent feature is a dense vertical 'slab' of galaxies near the centre of the surveyed volume. This feature is known as the Great Wall. A thin slice of the Great Wall accounts for the arms and shoulders of the Stick Man, but the Great Wall itself is an even larger feature, about 200 million light years long, almost perpendicular to the Stick Man.

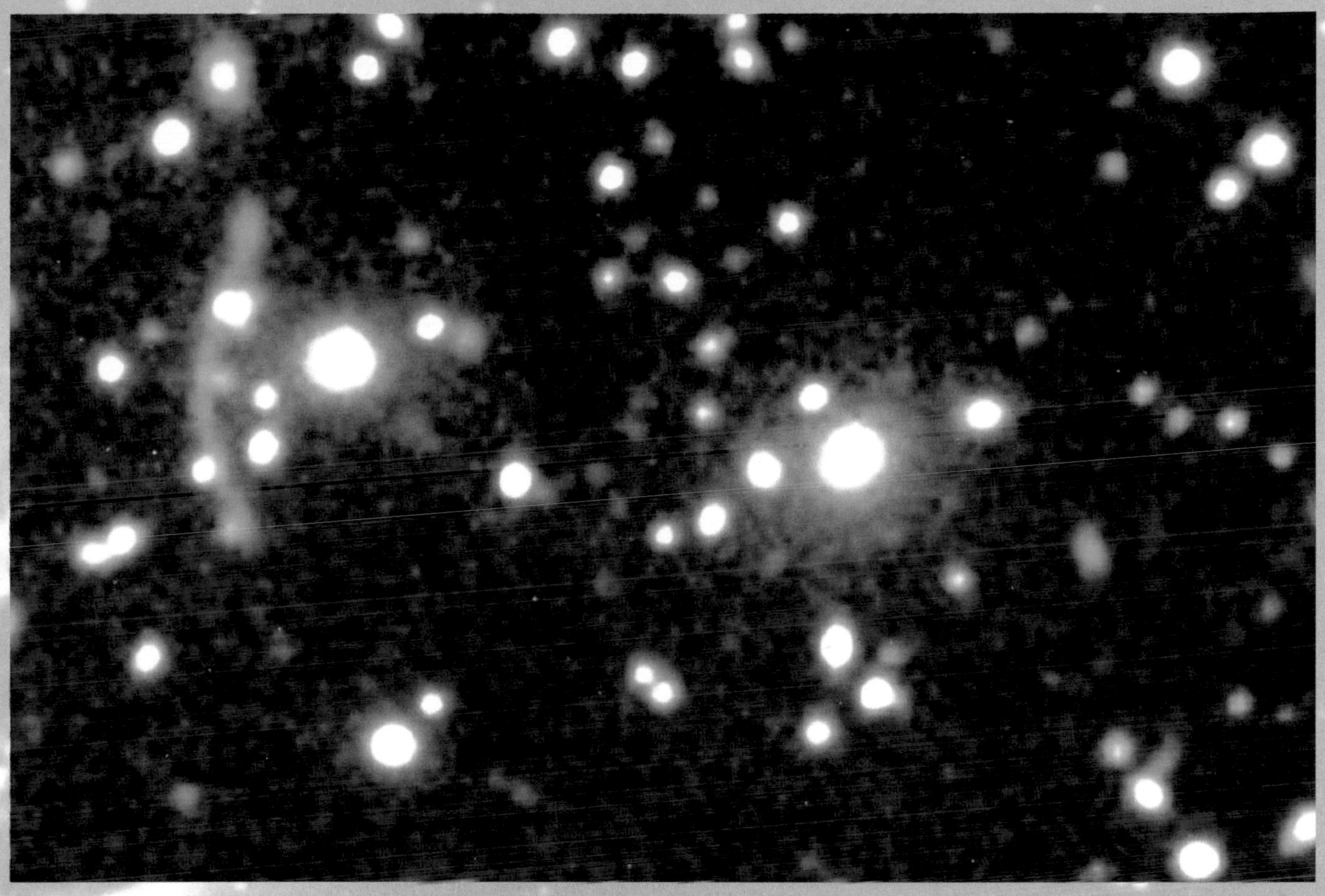

3.34 Gravitational lensing by a distant cluster of galaxies

Abell 370 is a rich cluster of galaxies at a great distance from the Local Supercluster. Like other clusters of galaxies, it is thought to contain a large amount of dark matter that is currently undetectable apart from its gravitational influence on visible matter. Within this image of Abell 370, a number of the cluster's galaxies can be seen as blue blobs. However, in addition there are a number of elongated blue arcs. These are believed to be distorted images of galaxies beyond Abell 370. The images are a result of the focusing of the light from the distant galaxies by the gravitational field of Abell 370. Such images can provide information about the mass of Abell 370 (including its dark matter) as well as providing a sort of natural telescope for the study of very distant galaxies. Many examples of this phenomenon of gravitational lensing are now known and there is little doubt that such 'lenses' will make a significant contribution to the study of galaxy clusters.

3.35 The expanding Universe: an analogy

Hubble's discovery, that the redshift of any sufficiently distant galaxy increases in rough proportion to the distance that separates it from us, shows that the Universe is expanding. The expanding surface of a gradually inflated balloon can mimic the presumed properties of this cosmic expansion in a number of ways. In the first place, the rate at which the distance between two points on the surface of the balloon increases will depend on the distance that separates those points. Secondly, although points on the balloon's surface move apart they do not move along the balloon's surface. Rather, they are carried along by the expansion of the balloon's fabric just as galaxies are thought to be carried along by the expansion *of* space rather than by motion *through* space. Finally, like the Universe, the *surface* of a balloon has no centre, and no point on the surface is more special than any other point.

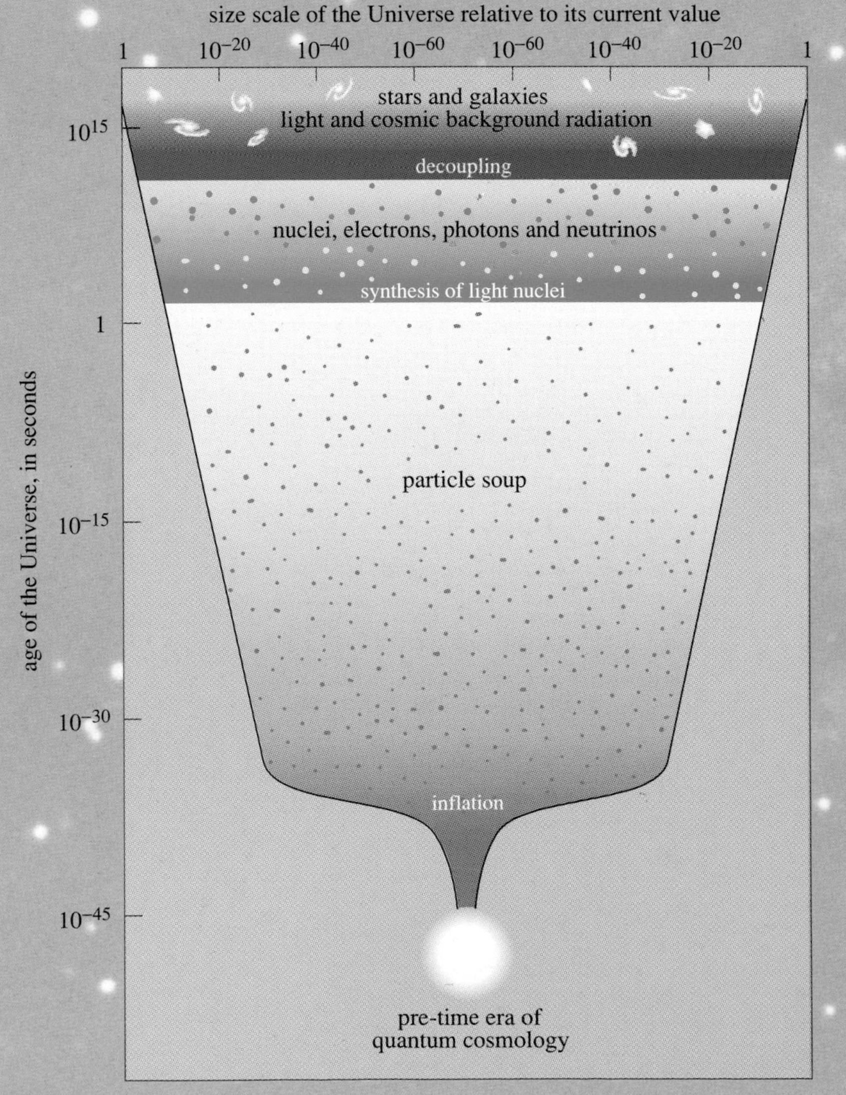

3.36 The history of the Universe

According to the Big Bang theory, our Universe has evolved over the past 10 to 20 thousand million years from a much hotter and denser state that was dominated by the radiation that it contained. The origin of this hot dense state is uncertain, but it is widely speculated that some sort of quantum fluctuation was initially responsible for the birth of the Universe and that this was followed by a period of enormously rapid expansion known as inflation. At the end of the inflation the Universe would have been a hot dense soup of elementary particles and radiation, expanding at a relatively sedate pace – just as the Big Bang theory requires. About three minutes later, as this cosmic soup cooled to temperatures below a thousand million degrees, low mass nuclei started to form, particularly helium nuclei. About 300 thousand years later the Universe was cool enough for matter and radiation to decouple as the first atoms formed. The result was an expanding cloud of gas, containing hydrogen and helium, but probably dominated by particles of dark matter. This expanding gas cloud went on to form the stars and galaxies we see today.

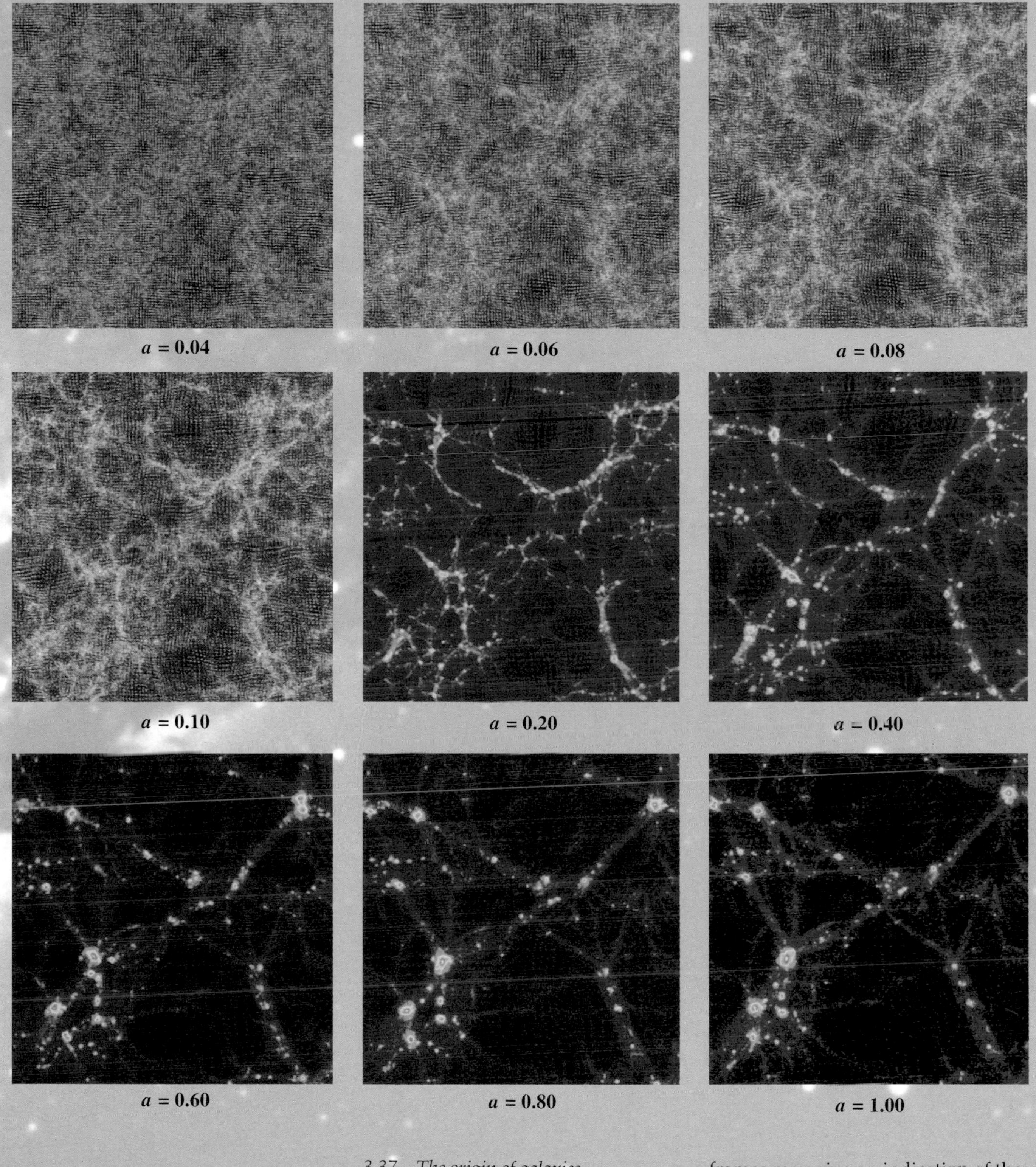

3.37 The origin of galaxies

These nine frames are snapshots from a computer simulation designed to show the evolution of the expanding cloud of atoms and dark matter particles that emerged from the Big Bang. The associated numbers show the size of the simulated volume relative to the final frame. The first few frames represent early epochs, before the galaxies formed, while the later frames may give an indication of the future evolution of the Universe. Galaxy formation is thought to have occurred mainly at the time simulated in the fifth frame; current conditions correspond to the seventh frame. If such simulations are correct, the galaxies we see today are located at those places where dark matter has most densely gathered itself together owing to the effects of its own gravitation.

(a)

(b)

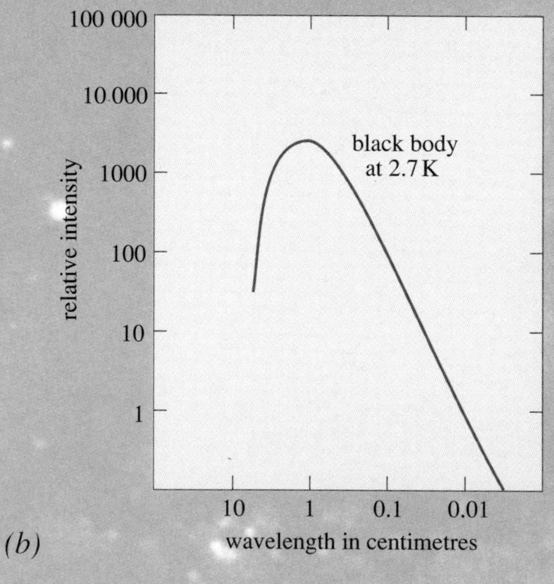

3.38 The discovery of the cosmic microwave background radiation

(a) In 1964, while this horn antenna was being used to investigate satellite communications, Arno Penzias and Robert Wilson discovered a 'background' of microwave radiation coming from all directions in space. After consultations with physicists from Princeton University, they realized that this was probably the massively redshifted remnant of the radiation that decoupled from matter when the first atoms formed and the Universe become transparent, 300 thousand years after the Big Bang. Their discovery, which won them a Nobel Prize, played a major part in establishing the Big Bang as the standard model of the origin and early evolution of the Universe.
(b) Subsequent work showed that the radiation has a spectrum (a distribution of relative intensity against wavelength) that is characteristic of an ideal radiation source with an absolute temperature of about 2.7 kelvins (about -270 degrees Celsius).

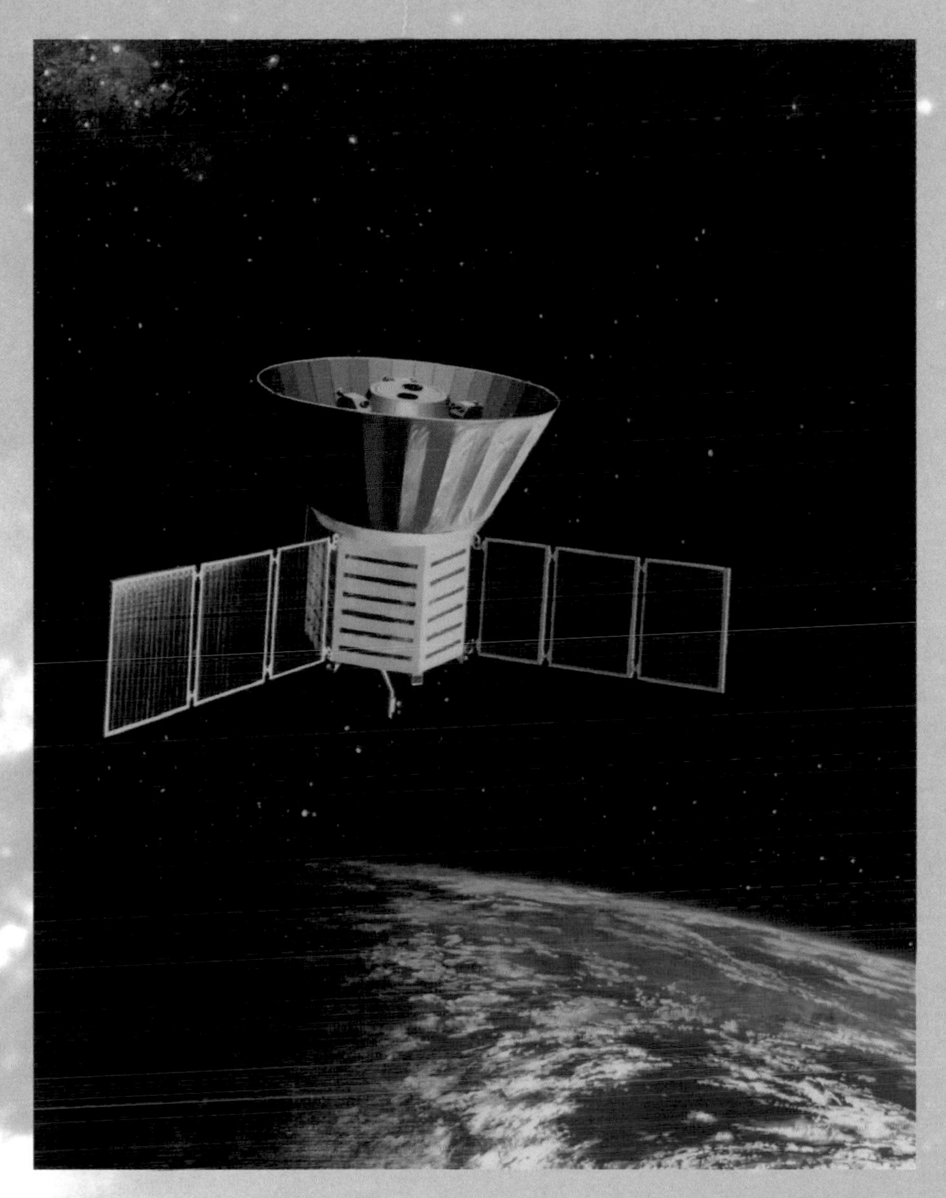

3.39 COBE: the Cosmic Background Explorer
The cosmic background radiation is of such importance that a satellite dedicated to its study was launched in 1989 . That satellite, the Cosmic Background Explorer, confirmed that the spectrum of the background radiation is indeed characteristic of an ideal source (at absolute temperature 2.7 kelvins) and thus verified a major prediction of the Big Bang. The open end of the cone is a few metres across.

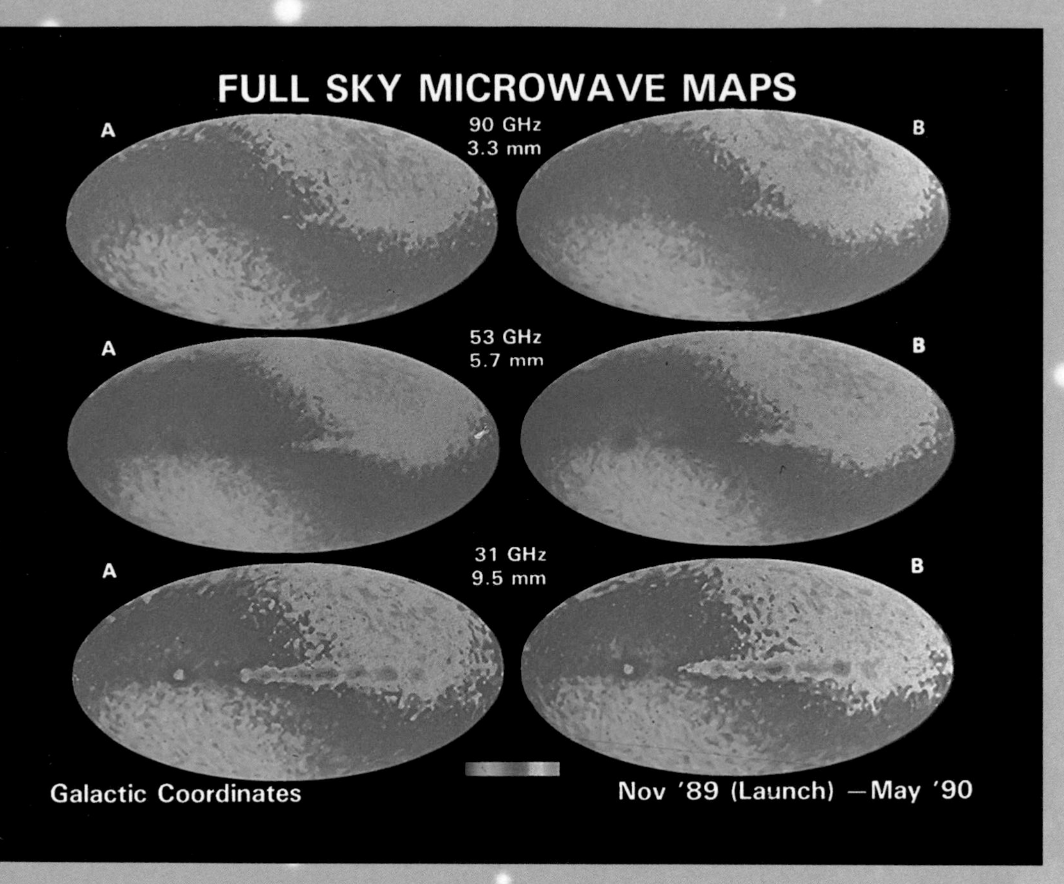

(a)

3.40 The uniformity of the cosmic background radiation

(a) These all-sky maps – plotted in galactic coordinates with the plane of the Milky Way running across the middle, and centred on the Galactic centre – show variations in the effective temperature of the background radiation. COBE has made such measurements at three different wavelengths using two different detector channels (marked A and B in the images). Those observations show that microwave radiation coming from one part of the sky (shown in pink) has a slightly higher temperature than that from the diametrically opposite part of the sky (blue). This variation is known as the dipole effect, and is a consequence of the Earth's motion relative to the overall expansion of the Universe. (b) Once the dipole effect has been subtracted from the data, the resulting maps show a highly uniform temperature distribution, as predicted by the Big Bang.

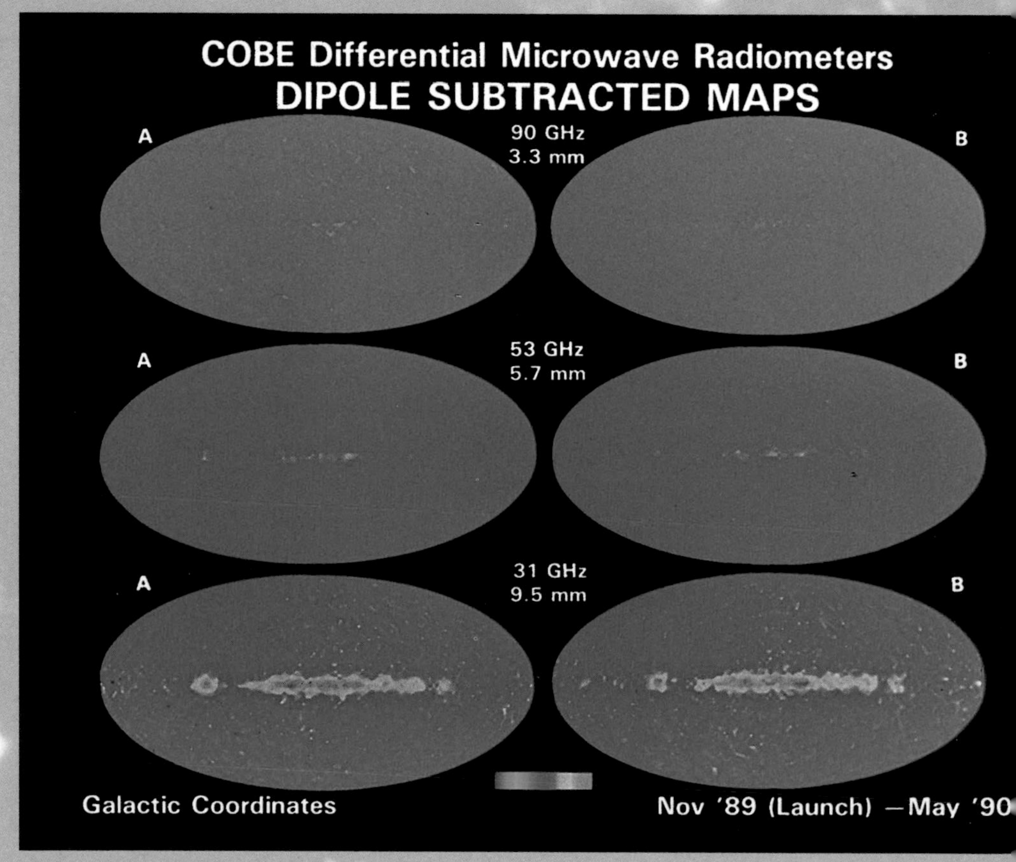

(b)

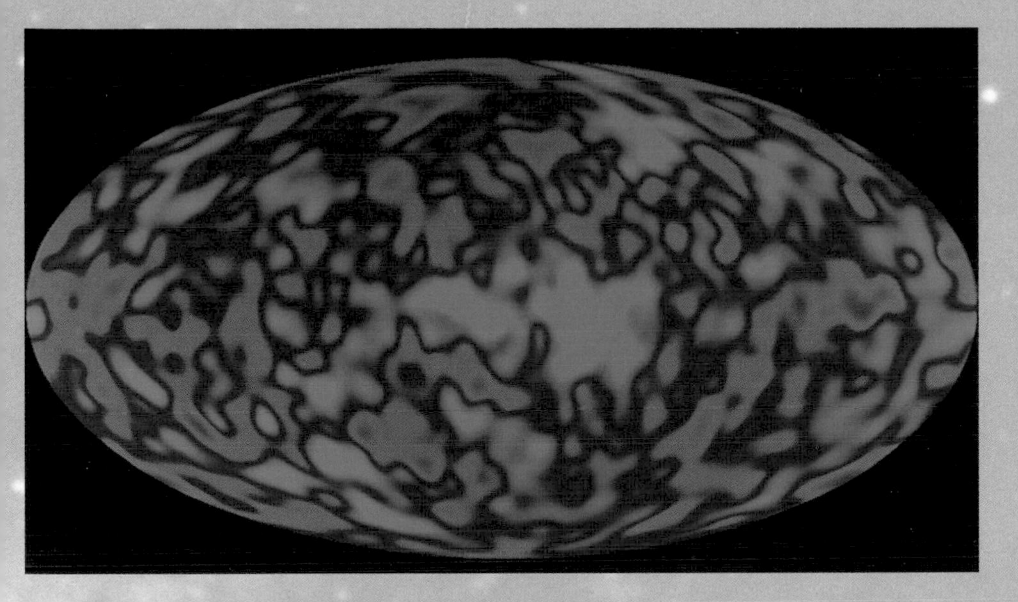

3.41 Departures from uniformity in the cosmic background radiation

Although the Big Bang theory predicts a *highly* uniform cosmic background radiation, that background should not be *perfectly* uniform. According to the theory, the properties of radiation and matter were closely related throughout the first 300 thousand years of the Universe's history. Thus, because galaxies are thought to have arisen from localized variations in the density of matter that were present from the very earliest times, there should be corresponding departures from uniformity in the temperature of the cosmic background radiation. COBE has detected such variations, though they are very tiny (just one part in a hundred thousand) and they are masked by random 'noise'. The discovery of these variations gives further support to the Big Bang theory and emphasizes the significance of the cosmic background radiation.

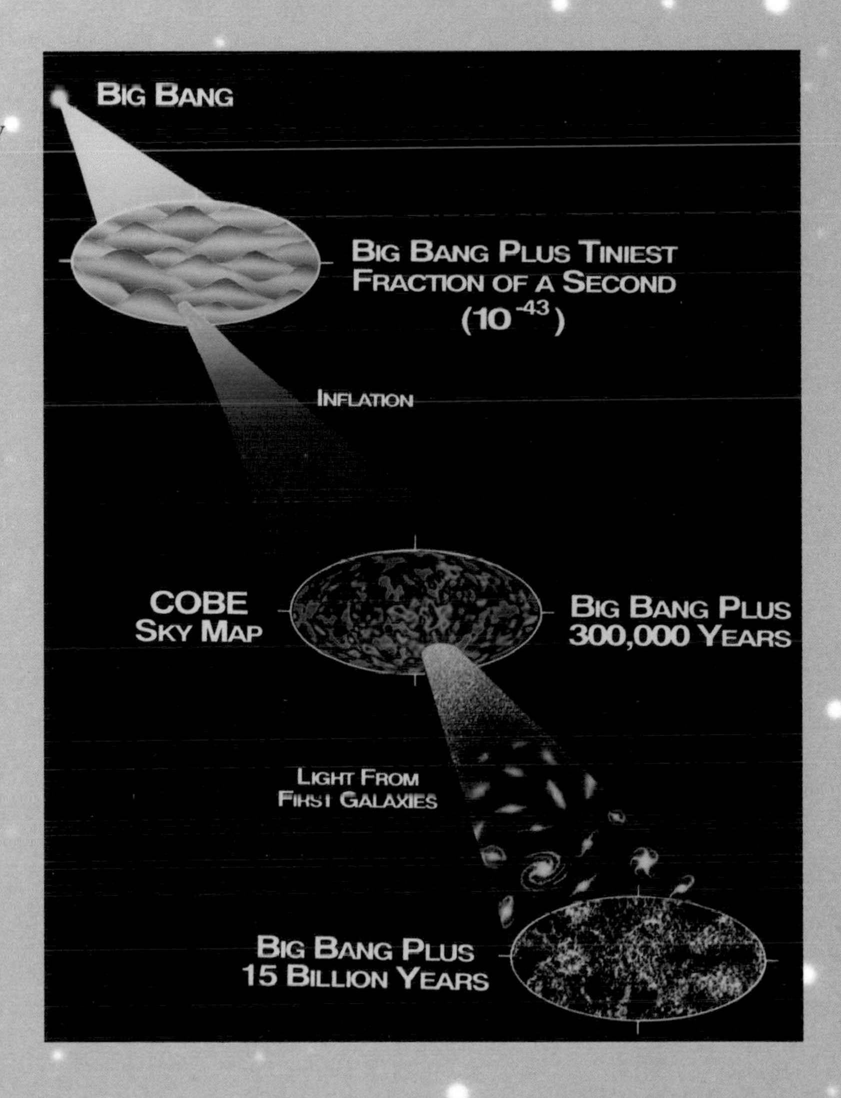

3.42 The evolution of cosmic structure

If the Universe was created by some sort of quantum process, as is widely assumed, then, even at the very earliest times, small variations in its density might well be expected. These variations would have been amplified by the period of ultra-rapid expansion known as inflation, and would have left their mark on the cosmic background radiation when it decoupled from matter about 300 thousand years after the Big Bang. The same variations would have also influenced the matter in the Universe and would have acted as the seeds of the galaxies and clusters of galaxies that we see today, about 15 thousand million years after the Big Bang. These closing images of the cosmos thus encapsulate the whole of cosmic history.

Maps of the sky

Maps of the sky use celestial latitude and longitude, which are analogous to terrestrial latitude and longitude. Celestial latitude is called declination (dec), and celestial longitude is known as right ascension (RA). Imagine the Earth covered with grid lines of RA and dec, and then imagine expanding this grid out towards the stars, and then attached to an imaginary sphere called the celestial sphere. The Earth's Equator becomes the celestial equator, with zero declination, and the north and south poles become the north and south celestial poles, with declinations of +90 degrees and −90 degrees respectively. Celestial longitude could also be measured in degrees, running from 0 to 360 degrees. In fact, it is measured in hours, from 0 hours to 24 hours, and so 1 hour is equal to 15 degrees. This apparently odd choice is due to the Earth's rotation, which causes an apparent rotation of the sky once every 24 hours.

To show the celestial sphere on a flat page we have projected it as shown overleaf. The view we then get from the inside gives us the sky map divided into three parts: the equatorial region , the north polar region, and the south polar region.

On the sky maps we have shown the positions of a couple of constellations and some of the brightest stars. We have also shown the locations of the objects in the plates in Parts 1 and 3 by putting the plate numbers at the celestial coordinates of the centre of the plate. In most cases the amount of sky covered by the plate is smaller that these labels, but in some cases it is much larger. We have not shown the locations of objects that change their coordinates significantly on a human time-scale. Thus, we have not shown any of the Solar System objects in Part 2, or the Sun. These objects move right around the celestial sphere about once a year.

At any moment, from any point on the Earth's surface, only half of the celestial sphere is above the horizon, and celestial objects will be visible (given clear skies) only if the Sun is below the horizon. To find out what can be seen in a dark sky from your particular terrestrial latitude at any particular date and time, you can use for your locality month-by-month sky maps or a device called a planisphere.

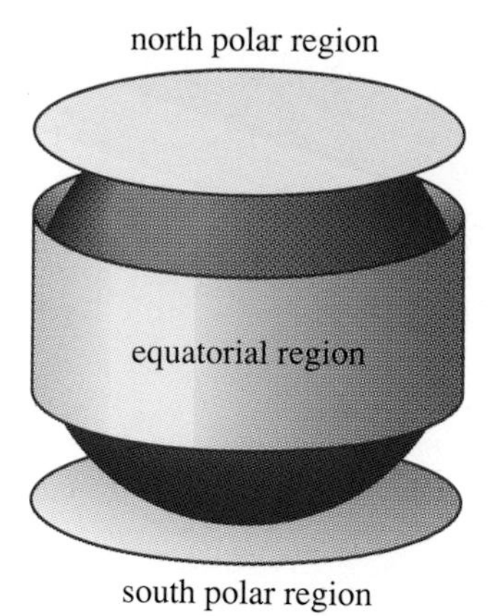

Projecting the celestial sphere on to a flat page.

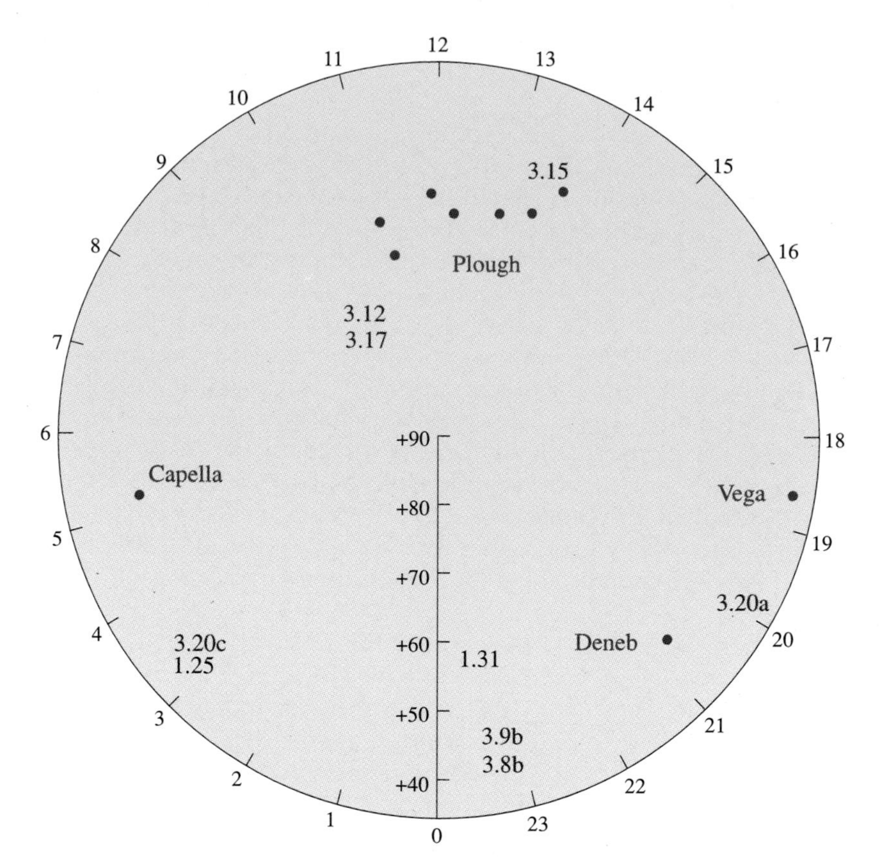

North polar region (2000.0).

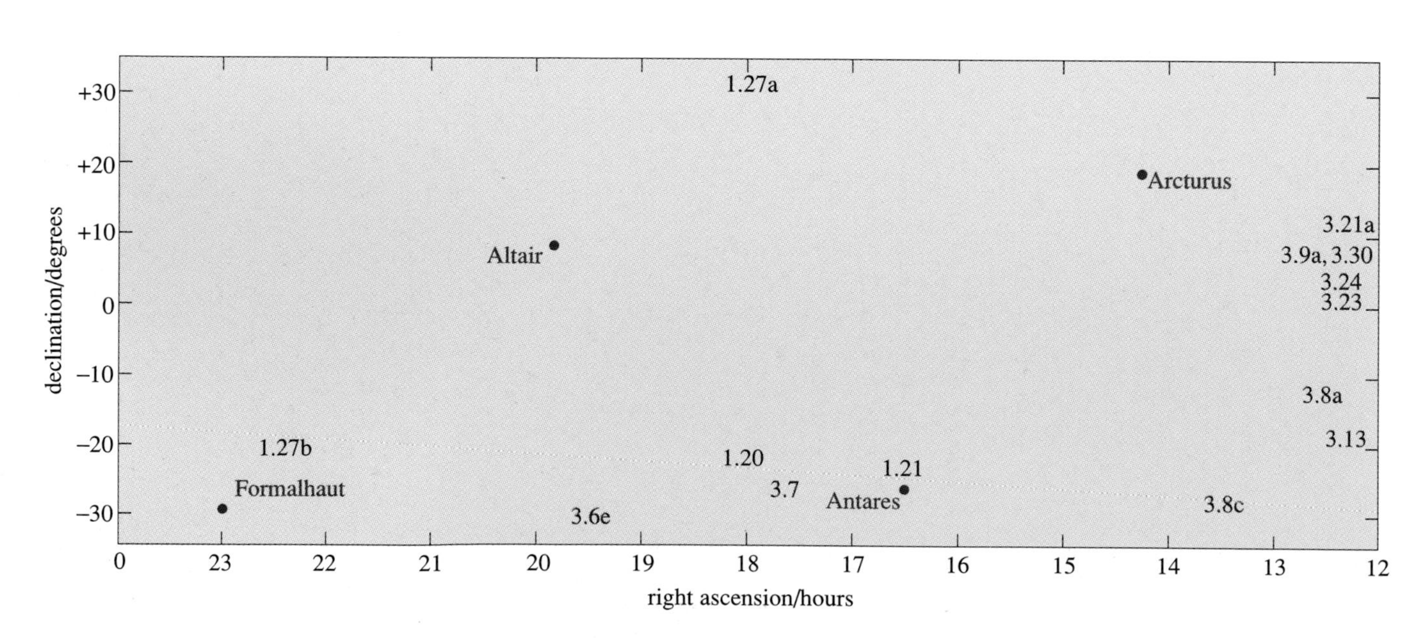

Equatorial region (2000.0).

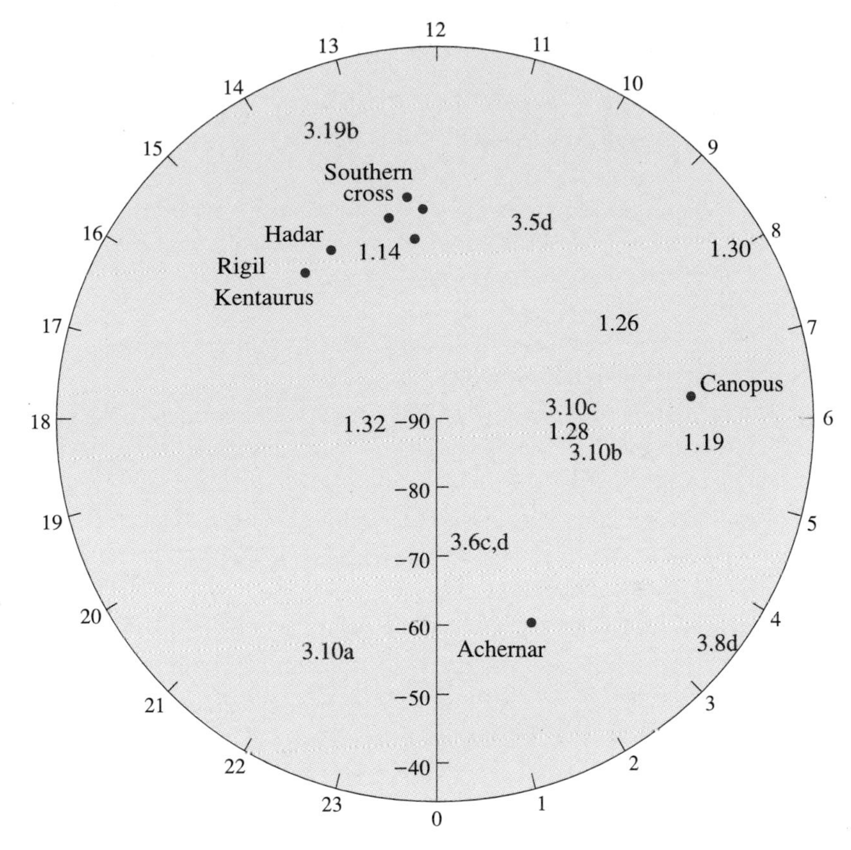

South polar region (2000.0).

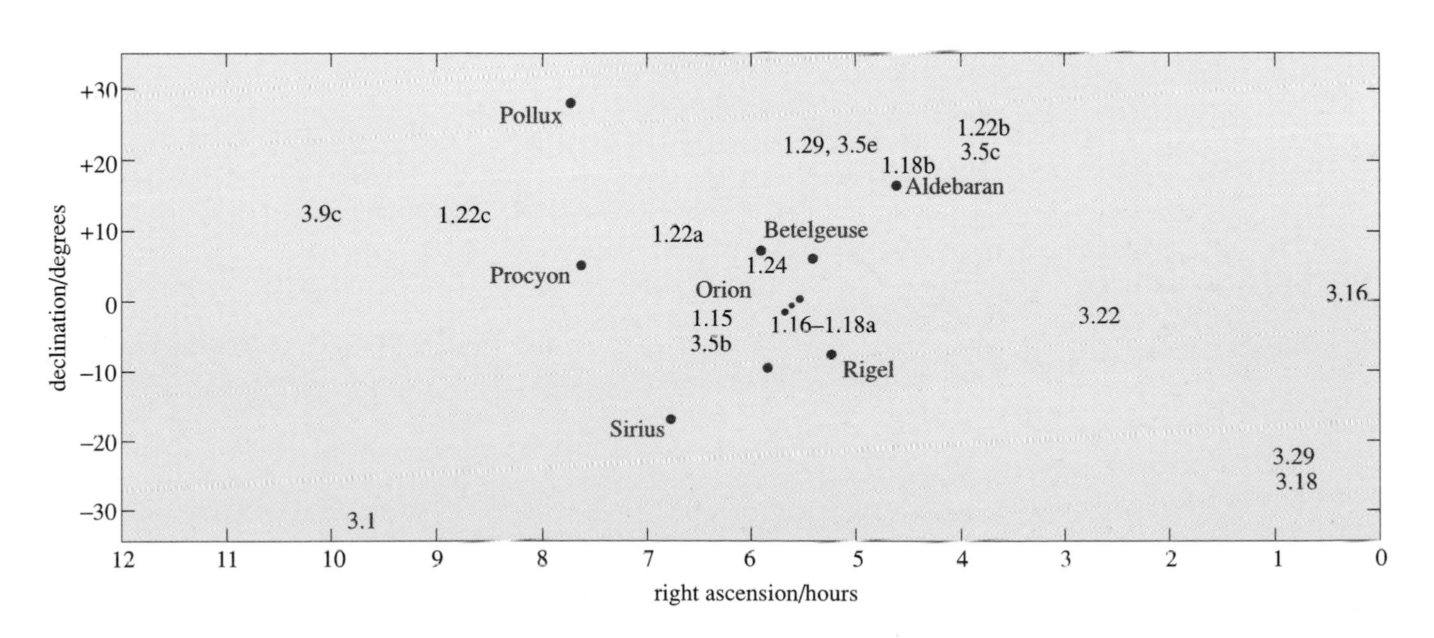

Glossary

Words printed in italics are explained elsewhere in this glossary.

absolute temperature A temperature measured on the absolute temperature scale, which starts at absolute zero, the lowest meaningful temperature (about –273 °C). Absolute temperatures are measured in a unit called the *kelvin*, which has the same size as the Celsius (centigrade) degree.

absorption line A feature of a *spectrum*, where, over a narrow range of *wavelengths* or *frequencies*, there is a dip in the radiation received.

accretion disc A disc of material orbiting a *black hole*, or some other dense object, and gradually feeding it with matter. Close to a black hole an accretion disc would be very hot and would be a source of X-rays.

active galaxy A *galaxy* with an unusually energetic nucleus. Active galactic nuclei can display variability on various time-scales and are often sources of radio waves or X-rays.

arcmin Abbreviation for a minute of arc. There are 60 arcmin in a degree of arc.

arcsec Abbreviation for a second of arc. There are 60 arcsec in a minute of arc (*arcmin*).

asteroids Irregular-shaped, mostly rocky, bodies ranging in size from nearly a thousand kilometres across down to large boulders. Nearly 5000 have been charted, mostly in *orbits* between those of Mars and Jupiter. If all the asteroids were assembled together, they would make a planet whose mass was less than one thousandth of the Earth's.

Big Bang theory A theory of the early evolution of the Universe that assumes that the present-day Universe has evolved from an initial state that was much hotter and denser, and which was dominated by the *electromagnetic radiation* that it contained. (The Universe is now dominated by matter, though electromagnetic radiation is still important.)

binary star Two *stars* in *orbit* around each other.

black hole A region of space in which a *star* or some other body has collapsed catastrophically under its own gravity. According to Einstein's general theory of relativity, no signal of any kind may emerge from within a black hole.

chondrite The most common kind of *meteorite*, consisting of rounded clusters of mineral crystals (known as *chondrules*), held together in a fine-grained rocky matrix.

chondrule A rounded cluster of mineral crystals that formed at a high temperature, and characteristic of chondritic *meteorites*.

comet A body often described as a 'dirty snowball', typically around 10 kilometres across, that follows an eccentric *orbit* around the *Sun*. As it approaches the Sun it heats up and some of the ice is vaporized. Both ice and dust then stream away from the Sun, and may produce a spectacular tail visible from Earth with the unaided eye.

cosmic background radiation Microwave radiation observed to come from all directions in space. The existence of cosmic background radiation is a prediction of the *Big Bang theory*.

cosmic rays High speed elementary particles, mainly protons, that stream through interstellar space. The origin of cosmic rays is uncertain, though many seem to originate within the *Milky Way*, perhaps in *supernovas*.

crater A near-circular depression (metres to hundreds of kilometres across) on the surface of a *planet* or a *planetary satellite*. Almost all appear to have been formed by impacts of Solar System debris (large *meteorites*, *asteroids* and *comets*). These can be distinguished from volcanic craters by the following features: floors lower than the surrounding terrain, raised rims, central peaks (except in small craters), fragmented material (*ejecta*) thrown out. The older a surface is, the more impact craters it has.

dark matter A non-luminous form of matter: its existence is deduced solely from its gravitational effect on more directly detectable forms of matter. The nature and distribution of dark matter are very uncertain. Many astronomers believe that the total mass of all the directly detectable matter in the Universe may be less than 10% of the mass of dark matter that is present. Many *galaxies* appear to be embedded in vast clouds of dark matter.

decoupling A process that occurred about 300 000 years after the *Big Bang*, when stable atoms (mainly hydrogen atoms) formed for the first time. Until that time *electromagnetic radiation* was strongly influenced by the behaviour of charged particles, such as atomic nuclei and electrons. Following the formation of electrically neutral atoms from electrons and nuclei, radiation ceased to be substantially influenced by the behaviour of matter.

dense cloud The densest, coolest type of region in the *interstellar medium. Stars* are born in dense clouds.

ecliptic plane The plane of the Earth's *orbit* round the Sun. The orbits of all the other *planets*, and of most *asteroids*, lie close to this plane.

ejecta Fragmented material that is thrown out when an impact *crater* is formed.

electromagnetic radiation A flow of energy consisting of radiation from all or part of the *electromagnetic spectrum*. (See also *electromagnetic waves*.)

electromagnetic spectrum A collective term used to describe the various *wavelength* ranges of *electromagnetic radiation*. In order of increasing wavelength, these ranges are: gamma rays, X-rays, ultraviolet waves, visible waves (light), infrared waves, microwaves, radio waves. (See also *spectrum*.)

electromagnetic waves A fluctuating pattern of electric and magnetic fields that travels through empty space at 300 000 kilometres per second. Such waves constitute *electromagnetic radiation*.

emission line A feature in a *spectrum*, where, over a narrow range of *wavelengths* or *frequencies*, there is a peak in the radiation received.

frequency The rate at which the peaks (or troughs) in a smooth repetitive wave pass a particular point in space as the wave travels along.

galaxy A large, gravitationally bound aggregate of *stars* and interstellar matter, possibly dominated by *dark matter*.

giant planet Any of the four largest *planets* in the Solar System (Jupiter, Saturn, Uranus, and Neptune), characterized by their immensely deep atmospheres, consisting largely of hydrogen.

Hadley cell Convection cell in a *planet's* atmosphere, notably Venus and the Earth, where warm air rises over the Equator, flows polewards, sinks after cooling, and flows back towards the Equator, where it gets heated again.

Hubble classification A classification scheme for *galaxies* based on their visual appearance: they are assigned to a Hubble type, such as elliptical, spiral or irregular.

Hubble's constant An important astronomical parameter that measures the rate of expansion of the Universe. Current estimates put Hubble's constant roughly in the range 14 to 30 kilometres per second per million light years. In this book a value of 23 kilometres per second per million light years has been assumed.

Hubble's law An approximate relationship between the distance of a *galaxy* and its *redshift*. According to Hubble's law, galactic redshifts increase in rough proportion to galactic distances.

icy satellite Several of the satellites of the *giant planets*, and also Pluto's satellite Charon, that are composed of a mixture of rock and ice. At Jupiter the ice is essentially frozen water, but farther from the *Sun* it includes more volatile frozen components, such as ammonia, methane and nitrogen. In all important respects these ices behave like rock does in the *terrestrial planets*.

image Any picture, however recorded. Most of the images in this book are not photographs. Photographic film is sensitive only in the visible and near-visible part of the *electromagnetic spectrum*, so electronic means have to be used to record an image outside this range. Even many of the visible images here were recorded electronically, notably those transmitted to Earth by spacecraft.

inclination The angle between the plane of the disc of a *galaxy* and the line of sight. Galaxies seen face-on are inclined at 90° to the line of sight.

inflation A hypothetical period of enormously rapid cosmic expansion that happened at a very early stage in the history of the Universe, according to some theories.

interstellar medium The tenuous medium between the *stars*, consisting of gas (mainly hydrogen) and a trace of dust. It is highly varied in density, temperature and chemical make-up, and so different types of region have different names, for example, *dense clouds* (where stars are born), *planetary nebulas* (shed by old stars with masses of less than about 8 times that of the Sun), and *supernova* remnants (created by stars of more than about 8 solar masses, when they explode at the ends of their lives).

kelvin (K) The unit of temperature on the *absolute temperature* scale.

lava Molten rock that has erupted at a *planet's* surface. Also used to refer to the fluids rich in water, ammonia or methane that can erupt on *icy satellites*.

light year (ly) The distance that *electromagnetic waves* travel through empty space in a year. These waves travel at the speed of light (300 000 kilometres per second), and cover 9.46 million million kilometres in a year.

lithosphere The rigid outer shell of a solid *planet* or *planetary satellite*, whether icy or rocky.

Local Group The sparse cluster of 20 or 30 galaxies that includes the *Milky Way*, the Magellanic Clouds and M31. The Local Group has a diameter of about 6 to 8 million light years.

Local Supercluster A cluster of clusters of galaxies with a diameter of about 100 to 150 million light years. The *Local Group* is an outlying member of the Local Supercluster.

magma As for *lava*, but referring to molten material at depth as well as at the surface.

main sequence star A *star* in the main sequence phase of its lifetime, when nuclear reactions in its central core are converting hydrogen into helium.

meteorite A lump of stone or nickel-iron alloy that falls to Earth. Most meteorites appear to be fragments of *asteroids* broken during collisions. A few have been found that are rocks from the Moon and Mars, presumably ejected into space by the impact of a small asteroid.

Milky Way Our own *galaxy*.

neutron star When a *supergiant* becomes a *supernova*, it blows off most of its mass, but can leave an extremely dense remnant, about 10 kilometres in diameter. This remnant, which consists mostly of neutrons, spins rapidly and can radiate a beam of radio waves. If the Earth happens to lie in the path of this beam a regular pulse of radio waves is observed, one pulse per rotation of the neutron star. The neutron star is then called a *pulsar*.

open cluster A cluster of several hundred *stars*, born in a *dense cloud*. Most open clusters gradually disperse to yield single stars, or small numbers close together in *orbit* around each other.

orbit The path followed by one body moving under the gravitational influence of another body. The Moon orbits the Earth, and the Earth orbits the Sun. *Binary stars* orbit each other. Orbits are elliptical in shape; those of *planets* around the Sun or of satellites around their planets are typically near-circular, but most *comets* have extremely elongated orbits.

planet One of the nine major bodies orbiting the Sun (Mercury, Venus, Earth, Mars, Jupiter, Saturn, Uranus, Neptune, and Pluto). The term planetary body is sometimes used in a general sense to include planets, asteroids and those satellites of the planets that are large enough to experience similar phenomena.

planetary nebula The shell of material cast off by a *star* that was earlier a *red giant*. It has nothing to do with planets!

planetary satellite A body orbiting a *planet*. (See also *icy satellite*)

pulsar See *neutron star*.

quasar A quasi-stellar object with a very large *redshift*. Although still controversial, it is widely assumed (including in this book) that quasars are a class of *active galaxies*. If this assumption is correct, quasars are amongst the most luminous of all *active galaxies* and represent some of the most distant astronomical objects ever observed.

radar The sending out of *electromagnetic radiation* in the microwave part of the *electromagnetic spectrum*, and the subsequent detection of some of this radiation after its reflection by an object. Radar is used to map planetary surfaces, and to explore the nature of the surfaces. Radar can penetrate atmospheric clouds.

radio galaxy A kind of *active galaxy* recognized by its characteristic radio emission.

red giant When a *main sequence star* with a mass of less than about 8 solar masses runs out of hydrogen nuclear fuel in its core, it swells up and its surface cools: it becomes a red giant. To the eye, red giants appear orange-white or yellowish-white.

redshift A fractional increase in the *wavelength* of a spectral feature, such as an *absorption line*, that results from an effect such as the motion of the source away from us or the expansion of the Universe.

solar activity The *Sun* goes through a cycle of activity once every 11 years or so. High activity is marked by higher sunspot number, by a greater frequency of local brightenings called solar flares, and of gaseous plumes called solar prominences. The solar wind, a gusty stream of particles ejected by the Sun, is more copious during high activity, and it can cause spectacular auroral displays in the Earth's upper atmosphere.

solar nebula A rotating cloud of gas and dust presumed to have developed within a *dense cloud* in the *interstellar medium* about 4600 million years ago, and within which the *Sun* and the rest of the bodies in the *Solar System* grew.

Solar System Our *Sun* and all the bodies associated with it (the *planets*, *planetary satellites*, *asteroids*, and *comets*).

spectrum A display of the strength of radiation versus *wavelength* or *frequency*. A spectrum can be continuous, i.e. smooth, or it can be marked by *emission lines* or *absorption lines*.

star A body that, on formation, becomes hot enough in a central core for nuclear reactions to occur that convert hydrogen into helium. During this core conversion the star is said to be in the *main sequence* phase of its lifetime, and it accounts for nearly all of the time between when the star began to form in a *dense cloud*, and when it becomes a *white dwarf* or a *supernova*.

Sun The star at the centre of our *Solar System*. The Sun is a *main sequence star* of modest mass, about halfway through its main sequence lifetime of about ten thousand million years.

supergiant When a *main sequence star* with a mass greater than about 8 solar masses runs out of hydrogen nuclear fuel in its core, it swells up and its surface cools: it becomes a supergiant. Supergiants are more luminous than *red giants*, and their surface temperatures can be higher, in which case they appear to the eye to be white, or bluish-white.

supernova An enormous explosion suffered by a *supergiant*, in which most or all of the star's mass is cast off, to form a supernova remnant in the *interstellar medium*. Any stellar remnant could become a *neutron star*, which we might observe as a *pulsar*. Alternatively, the stellar remnant might become a *black hole*.

terrestrial planet Any of the four innermost planets of the Solar System (Mercury, Venus, Earth, and Mars), characterized by rocky outer layers and metallic cores. The Earth's Moon and Jupiter's innermost large satellite Io have compositions and sizes comparable with these, and, from the point of view of the processes that can occur on and within them, can be regarded as terrestrial planets as well, even though they do not *orbit* the *Sun* independently.

wavelength The distance between adjacent peaks (or troughs) along a smooth repetitive wave.

white dwarf The small, Earth-sized, dense stellar remnant when a *star* sheds a *planetary nebula*. This is the final phase of the star's lifetime. At first a white dwarf is hot, and would appear bluish-white or white to the eye. But then it cools and eventually vanishes from view.

Tables

Some values

diameter of the Earth = 12 756 km
mass of the Earth = 5.98×10^{24} kg
diameter of the Sun = 1 392 000 km
mass of the Sun = 1.99×10^{30} kg
light year (ly) = 9.46×10^{12} km
60 arcsec = 1 arcmin
60 arcmin = 1 degree

Total eclipses of the Sun

Date	Maximum duration of totality/ minutes:seconds	Path of totality (a narrow strip passing through these regions)
3 November 1994	4:24	Peru, Chile, Bolivia, Paraguay, Brazil, South Atlantic
24 October 1995	2:10	Iran, Pakistan, India, Bangladesh, Burma, Cambodia, Vietnam, Pacific
9 March 1997	2:51	Mongolia, Siberia
26 February 1998	4:09	Pacific, Colombia, Panama, Venezuela, Atlantic
11 August 1999	2:23	North Atlantic, UK, France, Germany, Austria, Hungary, Rumania,Turkey, Iraq, Iran, Pakistan, India
21 June 2001	4:57	Atlantic, Angola, Zambia, Zimbabwe, Mozambique, Madagascar
4 December 2002	2:04	Angola, Botswana, Zimbabwe, South Africa, Mozambique, southern Indian Ocean, southern Australia
23 November 2003	1:59	Antarctica, southern Indian Ocean
8 April 2005	0:42	Pacific

Total eclipses of the Moon

Date	Maximum duration of totality/ minutes	From where visible (the whole, or nearly the whole, of these regions)
4 April 1996	86	Europe, Africa, South America
27 September 1996	70	North, Central and South America, West Africa
16 September 1997	62	Asia, Africa, Australasia
21 January 2000	76	North, Central and South America
16 July 2000	106	Eastern Asia, Australasia
9 January 2001	60	Asia, Africa, West Australia
16 May 2003	52	North, Central and South America
9 November 2003	22	Europe, West Africa, North, Central and South America
4 May 2004	76	Asia, Africa
28 October 2004	80	North, Central and South America, West Africa

The visually brightest stars

These are in descending order of visual brightness, as viewed from the Earth.

Name of star[a]	Distance/ light years	Coordinates (2000.0)[b] RA hr:min	dec deg:min	Type of star[c]	Spectral class[d]	Visual colour	Apparent visual magnitude[e]
Sun	0.000 015 8	–	–	MS	G2	yellowish-white	–26.73
Sirius (α CMa A)	8.6	06:45.1	–16:43	MS	A0/A1	bluish-white	–1.46
Canopus (α Car)	74	06:24.0	–52:42	G	A9	white	–0.72
Arcturus (α Boo)	34	14:15.7	+19:11	RG	K1.5	orange-white	–0.04
Rigil Kent[f] (α Cen A)	4.3	14:39.6	–60:50	MS	G2	yellowish-white	–0.01
Vega (α Lyr)	25	18:36.9	+38:47	MS	A0	bluish-white	0.03
Capella (α Aur AB)	41	05:16.7	+46:00	RG	G6 + G2	yellowish-white	0.08
Rigel (β Ori A)	910	05:14.5	–08:12	SG	B8	bluish-white	0.12
Procyon (α CMi A)	11	07:39.3	+05:14	MS/SG	F5	white	0.38
Achernar (α Eri)	69	01:37.7	–57:14	MS	B3	bluish-white	0.46
Betelgeuse (α Ori)	520	05:55.2	+07:24	SG	M2	orange-white	0.5
Hadar (β Cen AB)	320	14:03.8	–60:22	G	B1	bluish-white	0.6
Altair (α Aql)	16	19:50.8	+08:52	MS	A7	white	0.77
Aldebaran (α Tau A)	60	04:35.9	+16:31	RG	K5	orange-white	0.85
Antares (α Sco A)	326	16:29.4	–26:26	SG	M1.5	orange-white	0.9
Spica (α Vir)	220	13:25.2	–11:10	MS	B1	bluish-white	1.0
Pollux (β Gem)	35	07:45.3	+28:02	RG	K0	yellowish-white	1.14
Fomalhaut (α PsA)	22	22:57.6	–29:37	MS	A3	bluish-white	1.16
Mimosa (β Cru)	460	12:47.7	–59:41	G	B0.5	bluish-white	1.2
Deneb (α Cyg)	1500	20:41.4	+45:17	SG	A2	bluish-white	1.25
Acrux (α Cru A)	510	12:26.6	–63:06	sG	B0.5	bluish-white	1.33

[a] The name in brackets is the abbreviated formal name of the star. If the single letter A is added to the formal name then the star consists of several stars that, with the aid of a telescope, can be seen separately. The data then correspond to the brightest, A, component. If the two letters AB are added to the formal name then, even though the star is a binary, it cannot be seen as separate stars even in a telescope, and so for the purpose of this table it is treated as a single star.

[b] Referenced to the celestial coordinate system as it will be at the beginning of the year 2000. The coordinate system changes very slowly over many thousands of years.

[c] MS = main sequence, G = giant, RG = red giant, SG = supergiant, sG = subgiant.

[d] Spectral class is a measure of the surface temperature of the star.

[e] Apparent visual magnitude is a measure of visual brightness: the more positive the number the *dimmer* the star as it appears from the Earth. With the unaided eye we can see down to about 6. For variable stars the average magnitude is given.

[f] Full name, Rigil Kentaurus.

The nearest stars

These are in increasing order of distance from the Earth. Though nearby, most are too faint to be seen with the unaided eye.

Name of star[a]	Distance/ light years	Coordinates (2000.0) RA hr:min	dec deg:min	Type of star[b]	Spectral class[c]	Visual colour	Apparent visual magnitude[d]
Sun	0.000 015 8	–	–	MS	G2	yellowish-white	–26.73
Proxima Centauri[e]	4.2	14:30	–62:41	MS	M5.5	orange-white	11.05
α Centauri A[e]	4.3	14:40	–60:50	MS	G2	yellowish-white	–0.01
α Centauri B[e]	4.3	14:40	–60:50	MS	K1	yellowish-white	1.33
Barnard's Star	6.0	17:58	+04:34	MS	M3.8	orange-white	9.54
Wolf 359	7.7	10:56	+07:01	MS	M5.8	orange-white	13.53
BD+36°2147	8.2	11:03	+35:58	MS	M2.1	orange-white	7.50
L-726-8A[f]	8.4	01:39	–17:57	MS	M5.6	orange-white	12.52
L-726-8B[f]	8.4	01:39	–17:57	MS	M5.6	orange-white	13.02
α Canis Majoris A[g]	8.6	06:45	–16:43	MS	A1	bluish-white	–1.46
α Canis Majoris B[g]	8.6	06:45	–16:43	WD	–	bluish-white	8.3
Ross 154	9.4	18:50	–23:50	MS	M3.6	orange-white	10.45
Ross 248	10.4	23:42	+44:10	MS	M4.9	orange-white	12.29
ε Eridani	10.8	03:33	–09:28	MS	K2	yellowish-white	3.73
Ross 128	10.9	11:48	+00:48	MS	M4.1	orange-white	11.10
61 Cygni A[f]	11.1	21:07	+38:45	MS	K3.5	orange-white	5.22
61 Cygni B[f]	11.1	21:07	+38:45	MS	K4.7	orange-white	6.03
ε Indi	11.2	22:03	–56:47	MS	K3	orange-white	4.68
BD+43°44A[f]	11.2	00:18	+44:01	MS	M1.3	orange-white	8.08
BD+43°44B[f]	11.2	00:18	+44:01	MS	M3.8	orange-white	11.06

[a] These are the formal names or catalogue names of the stars. Few have popular names (see footnotes *e* and *g*).

[b] MS = main sequence, WD = white dwarf.

[c] Spectral class is a measure of the surface temperature of the star.

[d] Apparent visual magnitude is a measure of visual brightness: the more positive the number, the *dimmer* the star as it appears from the Earth. With the unaided eye we can see down to about 6. For variable stars the average magnitude is given.

[e] Three stars in orbit around each other, called Rigil Kentaurus as a group.

[f] Two stars in orbit around each other – a binary star.

[g] A binary star, also known as Sirius A and B, or as Sirius when seen as a single star in the sky.

Some interesting objects that are readily seen

Here are a few examples of some of the different sorts of interesting objects that lie beyond the Solar System, and that are visible to the unaided eye, though small binoculars (e.g. 7 × 40) give a better view, particularly if firmly supported. Note that some types of object, such as planetary nebulas and supernova remnants, all require a good telescope for a convincing view, so are not included.

Name	Coordinates (2000.0)		From where readily visible (latitudes)[a]
	RA hr:min	dec deg:min	
Star clusters (open)			
Pleiades	03:47.1	+24:08	North of 56° South
Praesepe	08:40.1	+20:00	North of 60° South
Jewel Box	12:53.6	−60:20	South of 20° North
Star clusters (globular)			
47 Tucanae	00:24.0	−72:04	South of 8° North
Omega Centauri	13:26.8	−47:18	South of 33° North
M13 in Hercules	16:41.7	+36:28	North of 44° South
Bright nebulas (with young star clusters)			
Orion Nebula[b]	05:35.3	−05:24	North of 75° South
Carina Nebula	10:45.1	−59:41	South of 20° North
Dense clouds			
Coal Sack	12:51	−63:00	South of 17° North
Galaxies			
M31 in Andromeda	00:42.8	+41:16	North of 39° South
Small Magellanic Cloud	00:52.6	−72:48.0	South of 28° North
Large Magellanic Cloud	05:23.6	−69:45.4	South of 10° North
The Milky Way	–	–	From anywhere

[a] For your latitude, a month by month sky map or a planisphere will enable you to find out the best times to view these and other objects.

[b] With binoculars, a dense cloud can also be seen.

A selection of large ground-based telescopes

Name	Size[a], etc.	Location	Latitude/degrees	Altitude/metres
Optical[b]				
VLT[c]	4 × 8.2 m thin, with AO	Cerro Paranal, Chile	−25	2640
Keck[c]	2 × 9.8 m segmented, with AO	Mauna Kea, Hawaii	+20	4150
BTA	6.0 m	Mt Pastukhov, Caucasus	+44	2100
Hale	5.1 m	Mt Palomar, USA	+33	1710
WHT	4.2 m	La Palma, Canary Isles	+29	2330
4-metre telescope	4.0 m	Cerro Tololo, Chile	−30	2220
AAT	3.9 m	Siding Spring, Australia	−31	1150
Mayall	3.8 m	Kitt Peak, USA	+32	2120
CFH	3.6 m	Mauna Kea, Hawaii	+20	4200
NTT	3.5 m thin, first with AO	La Silla, Chile	−29	2350
UK Schmidt	1.2 m; wide field surveys	Siding Spring, Australia	−31	1150
McMath	1.6 m solar telescope	Kitt Peak, USA	+31	2100
Infrared				
UKIRT	3.8 m	Mauna Kea, Hawaii	+20	4190
Gemini	8.1 m, one per hemisphere	Mauna Kea, Hawaii Cerro Pachon, Chile	+20 −30	4100 2730
Submillimetre[d]				
JCMT	15 m	Mauna Kea, Hawaii	+20	4100
SEST	15 m	La Silla, Chile	−29	2350
Radio[e]				
GMRT	30 × 45 m	Pune, India	+19	560
VLA	27 × 25 m	Socorro, USA	+34	800
Australia telescope	1 × 64 m + 6 × 22 m	Parkes + Narrabri, Australia	−33, −30	various
Westerbork telescope	12 × 25 m	Westerbork, Netherlands	+53	10
Merlin	1 × 76 m + six smaller	southern UK	+53	various
5-kilometre telescope	8 × 13 m	Cambridge, UK	+52	20
NRAO	3 × 26 m	Greenbank, USA	+38	820
Effelsburg	100 m; fully steerable	Bonn, Germany	+51	60
Arecibo	305 m; non-steerable	Arecibo, Puerto Rico	+18	500

[a] The size is the primary mirror or dish diameter, in metres (m). AO stands for 'active optics' – automatic optical adjustments that optimize the shape of the primary mirror and the position of the secondary mirror.

[b] 'Optical' means visible wavelengths and, in most cases, the infrared and ultraviolet wavelengths nearest the visible.

[c] These mirrors, with some smaller ones, can be used together to attain the performance of a larger single mirror.

[d] Submillimetre wavelengths are fractions of a millimetre, and thus encompass the very longest infrared wavelengths, and the very shortest microwavelengths.

[e] No distinction is made here between radiotelescopes that operate at microwavelengths, and those that operate at radio wavelengths. Note that single radiotelescopes can also be part of an array denoted, for example, by 30 × 45 m, and that some telescopes in an array can also be used singly. Also, some arrays can be incorporated into yet bigger arrays.

Important space telescopes

This is a selection of some of the world's important telescopes that are operating/have operated in orbit around the Earth. The Earth's atmosphere is opaque to most ultraviolet wavelengths, and to X-rays and gamma rays, and so space telescopes *have* to be used for these parts of the spectrum. At some infrared wavelengths the atmosphere is again opaque, and over most of the infrared it 'glows', making ground-based work difficult.

Name	Size[a], etc	Launch date	Termination date
COS-B	0.19 m^2 + smaller; gamma rays	August 1975	April 1982
CGRO	0.02 m^2; gamma rays	April 1991	continuing
Exosat	0.016 m^2; X-rays	May 1983	April 1986
Ginga	0.05 m^2; X-rays	February 1987	continuing
ROSAT	0.011 m^2 + smaller; X-rays	June 1990	continuing
Astro-D	0.13 m^2 + smaller; X-rays	February 1993	continuing
IUE	0.45 m; ultraviolet	January 1978	continuing
EUVE	0.5 m^2; ultraviolet	June 1992	continuing
Hubble	2.4 m; optical[b] + ultraviolet	April 1990	continuing
Hipparcos	0.3 m; optical; high precision positions	August 1989	continuing
IRAS	0.6 m; infrared	January 1983	December 1983
COBE	0.19 m; far infrared + submillimetre[c]	November 1989	continuing

[a] The size is either the collector diameter in metres (m), or its area in square metres (m^2).

[b] Optical wavelengths cover the visible, and the nearest infrared and ultraviolet wavelengths to the visible.

[c] Submillimetre wavelengths are fractions of a millimetre, and thus encompass the very longest infrared wavelengths, and the very shortest microwavelengths.

Basic data on the planets (including the Moon)

	Mercury	Venus	Earth	Moon	Mars	Jupiter	Saturn	Uranus	Neptune	Pluto
Average distance from Sun/10^6 km	57.9	108.2	149.6	149.6	227.9	778.3	1427	2870	4497	5900
Average distance from Sun/AU[a]	0.38	0.72	1	1	1.52	5.2	9.5	19.1	30.0	39.4
Orbital period	88.0 days	224.7 days	365.3 days	(27.3 days)	687 days	11.86 years	29.46 years	84.01 years	164.79 years	247.69 years
Axial rotation period/days	58.64	243.0	0.997	27.32	1.026	0.410	0.444	0.718	0.768	6.387
Axial inclination	0.0°	177.3°	23.4°	6.7°	25.2°	3.1°	26.7°	97.9°	29.6°	94°
Inclination of orbit to ecliptic plane[b]	7.0°	3.4°	0.0°	5.2°	1.8°	1.3°	2.5°	0.8°	1.8°	17.2°
Orbital eccentricity[c]	0.206	0.007	0.017	0.055	0.093	0.048	0.055	0.048	0.010	0.249
Equatorial radius/10^3 km	2.439	6.052	6.378	1.738	3.394	71.4	60.0	25.6	24.3	1.18
Mass/10^{24} kg	0.330	4.87	5.98	0.0735	0.643	1900	569	86.8	102	0.014
Mass/mass of Earth	0.0553	0.815	1.00	0.0123	0.107	317.8	95.2	14.5	17.2	0.002
Density/10^3 kg m^{-3}	5.43	5.25	5.52	3.34	3.95	1.33	0.69	1.29	1.64	2.1
Surface gravity/m s^{-2}	3.7	8.9	9.8	1.6	3.7	24.8[d]	10.6[d]	8.9[d]	11.6[d]	0.4
Satellites	0	0	1	–	2	16	18	15	8	1
Rings	0	0	0	0	0	few	many	several	few	0?
Mean surface temperature/°C	170	460	15	1	–50	–143[d]	–195[d]	–201[d]	–220[d]	–205 to –165[e]
Atmosphere (main components)	He, H	CO_2	N_2, O_2	Ne, Ar H_2, He	CO_2	H_2[d], He, CH_4	H_2[d], He, CH_4	H_2[d], He	H_2[d], He	CH_4, N_2?, CO?
Mean surface pressure of atmosphere/bars	10^{-15}	92	1.0	2×10^{-14}	0.006	n/a	n/a	n/a	n/a	variable

[a] AU stands for astronomical unit, and is equal to the average distance of the Earth from the Sun.

[b] The ecliptic plane is the plane of the Earth's orbit about the Sun, and is used as a reference against which to measure the inclination of other orbits.

[c] Eccentricity is a measure of the shape of an orbit. An orbit with an eccentricity of zero would be a perfect circle. The larger the value of eccentricity, the more elliptical the shape.

[d] The four giant planets (Jupiter, Saturn, Uranus and Neptune) are fluid to considerable depth; values of gravity, temperature and composition quoted here are for the atmospheric layer where the pressure is equal to the Earth's atmospheric pressure at sea-level (i.e. 1 bar).

[e] The average surface temperature of Pluto is not accurately known, but probably varies considerably as the planet moves around its eccentric orbit.

Planetary satellites

Planet	Satellite	Average distance from planet/10^3 km	Orbital period/days	Equatorial radius/km	Mass/ 10^{20} kg	Density/ 10^3 kg m^{-3}
Earth	Moon	384.5	27.32	1738	734.9	3.34
Mars	Phobos	9.4	0.319	14×10^a	1.3×10^{-4}	2.0
	Deimos	23.5	1.26	8×6^a	1.8×10^{-5}	1.7
Jupiter	Io	421.6	1.77	1815	894	3.57
	Europa	670.9	3.55	1569	480	2.97
	Ganymede	1070	7.16	2631	1482	1.94
	Callisto	1883	16.7	2400	1077	1.86
	12 others			$<135^a$		
Saturn	Mimas	186	0.942	197	0.38	1.17
	Enceladus	238	1.37	251	0.8	1.24
	Tethys	295	1.89	524	7.6	1.26
	Dione	377	2.74	559	10.5	1.44
	Rhea	527	4.52	764	24.9	1.33
	Titan	1222	15.94	2575	1346	1.88
	Iapetus	3561	79.3	718	18.8	1.21
	11 others			$<175^a$		
Uranus	Miranda	130	1.42	236	0.75	1.35
	Ariel	191	2.52	579	13.5	1.66
	Umbriel	266	4.15	586	12.7	1.51
	Titania	436	8.70	790	34.8	1.68
	Oberon	583	13.46	762	29.2	1.58
	10 others			$<85^a$		
Neptune	Proteus	178	1.12	209	0.4	1.1
	Triton	345	5.88	1350	214	2.08
	Nereid	5510	365.2	170	0.2?	1.2?
	5 others			$<100^a$		
Pluto	Charon	19.6	6.39	620	11	1.3

[a] These bodies are not spherical.

The four largest asteroids

Asteroid	Average distance from Sun/10^6 km	Average distance from Sun/AU[a]	Orbital period/ years	Inclination of orbit to ecliptic plane[a]	Orbital eccentricity[a]	Equatorial radius/km	Mass/ 10^{20} kg
Ceres	413.9	2.77	4.61	10.6°	0.097	457	≈10
Pallas	414.5	2.77	4.61	34.8°	0.180	261	≈2
Vesta	353.4	2.36	3.63	7.2°	0.097	250	≈2
Hygiea	470.3	3.14	5.59	3.84°	0.136	215	≈1

[a] For definitions, see the table of *Basic data on the planets.*

Selected comets

Comet	Perihelion[a] distance from Sun/10^6 km	Perihelion[a] distance from Sun/AU[b]	Orbital period/years	Inclination of orbit to ecliptic plane[b]	Orbital eccentricity[b]
Bennett	80.5	0.538	≫1000	90.0°	0.996
Encke	51.0	0.341	3.31	11.9°	0.846
Giacobini–Zinner	154.7	1.034	6.61	31.8°	0.706
Grigg–Skjellerup	148.8	0.995	5.10	21.1°	0.664
Halley	87.8	0.587	76.0	162.2°	0.967
Kohoutek	21.2	0.142	≫1000	14.3°	1.000[c]
Swift–Tuttle	144.1	0.963	120	113.6°	0.960

[a] The term perihelion means the closest point to the Sun reached during an orbit.

[b] For definitions, see the table of *Basic data on the planets.*

[c] An eccentricity of 1.0 means that the orbit is a parabola (which is an infinitely long ellipse).

Selected meteor streams

Meteor stream[a]	Date	Hourly rate[b]	Associated comet[c]
Quadrantids	3 January	30	not known
Eta Aquarids	4 May	10	Halley
Arietids	8 June	40	not known
Zeta Perseids	9 June	30	not known
Beta Taurids	30 June	20	Encke
Delta Aquarids	30 July	15	not known
Perseids	12 August	40	Swift–Tuttle
Draconids	10 October	–	Giacobini–Zinner
Orionids	21 October	15	Halley
Geminids	13 December	38	1983TB
Ursids	22 December	38	Tuttle

[a] Meteor streams are named after the region of the sky from which (from Earth perspective) they appear to radiate.

[b] This is the maximum number of meteors belonging to the stream visible per hour to the naked eye in a totally dark clear sky on the date specified. In practice you are likely to spot fewer than this.

[c] Most meteor streams are strung out along the orbits of known comets.

Selected missions of planetary exploration

Name	Launch date	Encounter date	Target	Notes	Operator
Luna 3	1959	1959	Moon	First pictures of the lunar far side	USSR
Surveyor 1	1966	1966	Moon	Soft landing and pictures	USA
Lunar Orbiter 1–5	1966–67	1966–67	Moon	Unmanned orbital photography	USA
Apollo 11, 12, 14–17	1969–72	1969–72	Moon	Manned landings, surface experiments and sample return	USA
Mariner 9	1971	1971–72	Mars	Orbiter (images and other data)	USA
Pioneer 10	1972	1972	Jupiter	First probe to cross the asteroid belt. Crude images of Jupiter, magnetic field and radiation measurements	USA
Mariner 10	1973	1974–75	Mercury	Three flybys by a single probe; detailed images of about half the planet	USA
Venera 9	1975	1975	Venus	Soft landing, chemical analyses and images transmitted to Earth	USSR
Viking 1	1975	1976	Mars	Detailed imaging from orbiter, lander provided surface images and experimental data	USA
Pioneer 12	1978	1978–79	Venus	Orbiter collecting radar altimetry data	USA
Voyager 1	1977	1979	Jupiter and satellites	Flyby (images and other data)	USA
		1980	Saturn and satellites	Flyby (images and other data)	
Voyager 2	1977	1979	Jupiter and satellites	Flyby (images and other data)	USA
		1981	Saturn and satellites	Flyby (images and other data)	
		1986	Uranus and satellites	Flyby (images and other data)	
		1989	Neptune and satellites	Flyby (images and other data)	
Venera 15, 16	1983	1983	Venus	Radar images of northern hemisphere from orbit (1-2 km resolution)	USSR
Vega 1	1984	1985	Venus	Atmospheric balloon, surface analyses	USSR
		1986	comet Halley	Complementary, more distant, images and analyses to those obtained by Giotto	
Giotto	1985	1986	comet Halley	Images, chemical and physical analyses	Europe
		1992	comet Grigg–Skjellerup	Chemical and physical analyses	
Magellan	1989	1990–93	Venus	Radar images of whole planet (≈120 m resolution)	USA
Galileo	1989	1991	asteroid Gaspra	First detailed images of an asteroid	USA
		1993	asteroid Ida	Second imaging of an asteroid	
		1995	Jupiter and satellites	First probe to orbit Jupiter. Imaging of atmosphere and satellites, entry probe to descend into atmosphere	
Mars Observer	1992	1993	Mars	Communication lost on arrival at Mars	USA
Mars 94	1994	1995–97	Mars	Stereoscopic imaging from orbit, atmospheric balloon rover	Russia–Europe
Cassini	due 1996	due 2002	Saturn and satellites	First probe to orbit Saturn. Imaging of atmosphere and satellites. Imaging radar to map Titan, and entry probe (Huygens) to descend through Titan's atmosphere	USA–Europe
Pluto Fast Flyby Mission	1999 or later	2006 or later	Pluto and Charon	Flyby. First detailed imaging of Pluto and its satellite.	USA

Probable members of the Local Group of galaxies

Name	Distance/10^6 ly	Hubble type	Coordinates (2000.0)		Diameter/10^3 ly	Apparent visual magnitude
			RA hr:min	dec deg:min		
Milky Way	–	Sbc	–	–	100	–
LMC	0.2	Ir	05:23.6	−69:47	32	0.1
SMC	0.2	Ir	00:52.7	−72:54	25	2.3
Ursa Minor	0.3	dE5	15:08.8	+67:07	2	12
Draco	0.3	dE3	17:20.2	+57:55	3	11
Sculptor	0.3	dE3	00:59.9	−33:42	3	10
Sextans I	0.5	dE	10:12.8	−01:41	3	9
Carina	0.5	dE4	06:41.7	−50:58	0.5	–
Fornax	0.5	dE3	02:39.6	−34:31	7	8
Leo I	0.8	dE3	10:08.5	+12:18	2	9.8
Leo II	0.8	dE0	11:13.5	+22:10	3	11.5
NGC 6822	1.7	Ir	19:44.9	−14:46	8	9
WLM	2.0	Ir	00:02.0	−15:28	7	10.9
NGC 147	2.4	E5	00:33.1	+48:31	10	9.3
NGC 185	2.4	dE3	00:38.9	+48:20	6	9.2
NGC 205	2.4	E5	00:40.3	+41:41	10	8.0
M31 = NGC 224	2.4	Sb	00:42.7	+41:16	200	3.4
M32 = NGC 221	2.4	E2	00:42.7	+40:52	5	8.2
Andromeda I	2.4	dE3	00:45.7	+38:00	2	13.2
Andromeda II	2.4	dE2	01:16.3	+33:25	2	13
Andromeda III	2.4	dE5	00:35.3	+36:31	3	13
IC1613	2.5	Ir	01:04.9	+02:07	12	9.3
M33 = NGC 598	2.5	Sc	01:33.9	+30:39	45	5.7
Pisces	3.0	Ir	01:03.7	+22:03	0.5	15
DD0210	3.0	Ir	20:47.0	−12:51	4	15
IC5152	3.6	Ir	22:02.9	−51:17	5	11
IC10	4.0	Ir	00:20.3	+59:19	6	10.3
GR8	4.0	Ir	12:59.2	+14:09	0.2	15
Sag DIG	4.0	Ir	19:30.0	−17:41	5	16

Acknowledgements

Grateful acknowledgement is made to the following sources for permission to reproduce illustrations in this book:

Part 1

Plate 1.1 Dr J. Dürst, Schönenberg, Switzerland; *Plate 1.2* George East; *Plate 1.3* National Optical Astronomy Observatories/NSO, Sacramento Peak; *Plate 1.4* from *Discovering the Universe* by W.J. Kaufmann, published by W.H. Freeman & Co, 1987. Reprinted with permission; *Plates 1.6, 1.9 and 1.11a* National Solar Observatory/Sacramento Peak; *Plates 1.7a, 1.15b, 1.32 and 1.33b* SERC/ Rutherford Appleton Laboratory, Didcot; *Plates 1.7b, 1.17c and 1.33d* NRAO/AUI; *Plates 1.7c, 1.22c, 1.27a and 1.29a* National Optical Astronomy Observatories; *Plate 1.7d* NASA/JPL. Image courtesy of Armagh Planetarium; *Plate 1.7e* Celestron International; *Plate 1.8* US Department of Navy, Naval Research Laboratory; *Plate 1.10a* D.H. Rooks, University College, London; *Plates 1.10b, 1.16 and 1.20* NASA; *Plate 1.11b* Big Bear Solar Observatory, California Institute of Technology; *Plate 1.12* Painting from *The New Solar System*, J. Kelly Beatty and Andrew Chaikin (editors) 1990, Sky Publishing Corp. Reproduced with permission; *Plate 1.13* NASA. Image courtesy of BDM International; *Plate 1.14* Akira Fujii, Tokyo; *Plates 1.15a, 1.17a, 1.17b, 1.22a, 1.26, 1.27b and 1.28a* Anglo Australian Telescope Board. Photographs by David Malin; *Plate 1.15c* R. Maddalena (NRAO), M. Morris, J. Moscowitz and P. Thaddeus; *Plate 1.18a* C. R. O'Dell (Rice University) and NASA; *Plate 1.18b* Ronald Snell, University of Massachusetts; *Plate 1.19a* D.A. Golimowski and S.T. Durrance (John Hopkins University) and M. Clampin, Space Telescope Science Institute, Baltimore; *Plate 1.19b* Dana Berry, Space Telescope Science Institute, Baltimore; *Plate 1.21a* The IRAS data were processed (or obtained) using the facilities of IPAC. IPAC is funded by NASA as part of the IRAS extended mission program under contract to JPL. Image courtesy of RAL; *Plates 1.21b, 1.22b and 1.30a* Royal Observatory Edinburgh/Anglo Australian Telescope Board. Photographs by David Malin; *Plates 1.23 and 1.25b* Julian Baum; *Plate 1.24* P. Warner, MRAO, Cavendish Laboratories, Cambridge, and William Hershel Telescope, La Palma; *Plate 1.25a* Painting by Steven Simpson. Reproduced with permission of *Sky & Telescope* magazine; *Plate 1.28b* NASA/ ESA. Image courtesy of Space Telescope Science Institute, Baltimore; *Plate 1.29b* from *Universe* by M. Rowan-Robinson published by Longman Group UK Ltd 1990; *Plates 1.30b, 1.31b and 1.33c* Max-Planck Institut für Extraterrestrische Physik, Germany; *Plate 1.30c* The IRAS data were processed (or obtained) using the facilities of IPAC. IPAC is funded by NASA as part of the IRAS extended mission program under contract to JPL; *Plate 1.31a* California Institute of Technology; *Plate 1.31c* Timothy Cornwell, Robert Duquet and NRAO/AUI; *Plate 1.33a* European Southern Observatory.

Part 2

Plates 2.1 and 2.2 Julian Baum; *Plates 2.3, 2.8, 2.9, 2.11, 2.26, 2.27, 2.33, 2.34b, 2.34c, 2.35a, 2.35b, 2.40, 2.41 and 2.51* Dr Bradford A Smith, National Space Science Data Center, World Data Center A for Rockets and Satellites, NASA Goddard Space Flight Center, Greenbelt, Maryland; *Plates 2.3 (inset), 2.4, 2.16, 2.17 (inset), 2.29 and 2.30* NASA; *Plates 2.5, 2.6 and 2.7* Mr Frederick J. Doyle, National Space Science Data Center; *Plates 2.10, 2.12, 2.13, 2.14, 2.15, 2.28 and 2.32* US Geological Survey, Flagstaff, Arizona; *Plates 2.17, 2.18, 2.19 and 2.20* Gordon H. Pettengill, the Magellan Project and the National Space Science Data Center; *Plates 2.21, 2.22 and 2.23* USGS Branch of Astrogeology, Flagstaff, Arizona; *Plate 2.24* processed by P.W. Francis in the Image Processing Laboratory at the Lunar and Planetary Institute. NASA Grant No 1929; *Plate 2.25* J. D. Griggs, USGS; *Plates 2.31 and 2.44* Dr Michael H. Carr, National Space Science Data Center;

Plate 2.34a NASA. Image supplied by Space Telescope Science Institute Baltimore; *Plates 2.34d, 2.37, 2.38 and 2.39* NASA/JPL. Images supplied by Armagh Planetarium; *Plate 2.36* G.W. Garneau, Jet Propulsion Laboratory, California Institute of Technology; *Plate 2.42* NASA/JPL. Image supplied by Space Telescope Science Institute, Baltimore; *Plate 2.43* Dr Michael J. S. Belton, National Space Science Data Center; *Plate 2.45* Gerhard Schwem, European Space Agency, Paris; *Plate 2.45 (inset)* ESA; *Plate 2.46a* Dr M. M. Grady, Natural History Museum, London; *Plate 2.46b* Dr I. A. Franchi, The Open University; *Plate 2.46c* Dr R. Hutchinson, Natural History Museum, London; *Plates 2.47 a–d and Plates 2.48 a–c* Natural History Museum, London; *Plate 2.49* D. Brownlee, University of Washington; *Plate 2.50* Johnson Space Centre.

Part 3

Plates 3.1, 3.8c, 3.9c, 3.10a, 3.18, 3.19b and 3.21a Anglo Australian Telescope Board. Photographs by David Malin; *Plate 3.2* Paul Doherty; *Plate 3.3* Dennis di Cicco, Sky Publishing Corp.; *Plate 3.4a* Glyn Haslam et al., Max Planck Institut für Radioastronomie, Bonn; *Plates 3.4b, 3.20a, 3.39, 3.41 and 3.42* NASA Goddard Space Flight Center, Greenbelt, Maryland; *Plate 3.4c* Lund Observatory, Sweden; *Plate 3.4d* Max Planck Institut für Extraterrestrische Physik, Germany; *Plates 3.5b, 3.5c, 3.10b, 3.10c and 3.13* Royal Observatory Edinburgh/Anglo Australian Telescope Board. Photographs by David Malin; *Plate 3.5d* Cerro Tololo Interamerican Observatory, AURA Inc. NOAO; *Plates 3.5e and 3.22* Dr Rudy Schild, Harvard Smithsonian Centre for Astrophysics, Washington; *Plate 3.6b* US Naval Observatory, Washington DC; *Plates 3.6c, 3.6e, 3.8a and 3.8d* European Southern Observatory; *Plate 3.6d* NASA/ESA. Image courtesy of Space Telescope Science Institute, Baltimore; *Plate 3.7a* from Order or Chaos, Chapter vi *The Milky Way,* by Ludwig Kuhn. Reprinted by permission of John Wiley & Sons Ltd, 1982; *Plate 3.7b* the IRAS data were processed (or obtained) using the facilities of IPAC. IPAC is funded by NASA as part of the IRAS extended mission program under contract to JPL; *Plates 3.7c, 3.19c, 3.20c and 3.21c* NRAO/AUI; *Plate 3.8b* Palomar Observatory, California Institute of Technology. Image courtesy of the Smithsonian Institute, Washington; *Plate 3.9a, 3.17 and 3.34* National Optical Astronomy Observatories; *Plates 3.9b and 3.21b* NASA. Images courtesy of Space Telescope Science Institute, Baltimore; *Plate 3.11a* Julian Baum, from an original by David Parker; *Plate 3.11b* Carnegie Institution, Washington; *Plate 3.12* Westerbork Synthesis Radio Telescope, Netherlands; *Plate 3.14* from Alar Toomre (MIT) and Juri Toomre (University of Colorado); *Plate 3.15* The Regents, University of Hawaii; *Plate 3.16* Durham University and the Royal Greenwich Observatory; *Plate 3.19a* from Peculiar Galaxies, Chapter 5 in *Foundations of Astronomy* by Michael A. Seeds, Wadsworth Publishing Co. Inc., 1988 (2nd ed.); *Plate 3.20b* Hale Observatories, California Institute of Technology; *Plate 3.23* Dane Penland, from J. Patrick Henry, Harvard Smithsonian Centre for Astrophysics, Washington; *Plate 3.24* Walter Jaffe/Leider Observatory, Holland. Ford/JHU/STScI and NASA; *Plate 3.25* from *Le Grand Atlas de l'Astronomie*, Encyclopedia Universalis, 1985; *Plate 3.26* illustration by George Retseck from The Quasar 3C 273 by J. Thierry, L. Courvoisier and E. I. Robson. © Scientific American June 1991. All rights reserved; *Plates 3.27 and 3.28* paintings by Steven Simpson. Reproduced with permission of *Sky & Telescope* Magazine; *Plate 3.29* S. J. Maddox, G. Efstathiou and W. J. Sutherland, University of Oxford; *Plate 3.30* Kitt Peak National Observatory, AURA Inc. Image courtesy of the Smithsonian Institute, Washington; *Plate 3.31* courtesy of R. Brent Tully, University of Hawaii; *Plate 3.32* adapted from material provided from V. de Lapparent, M. Geller and J. Huchra, Smithsonian Astrophysical Observatory; *Plate 3.33* Margaret J. Geller and John P. Huchra, Smithsonian Astrophysical Observatory; *Plate 3.37* James M. Gelb, NASA/Fermilab Astrophysics Centre; *Plate 3.38a* AT&T Bell Laboratories; *Plates 3.40a and 3.40b* NASA Goddard Space Flight Centre. Images courtesy of the Astronomical Society of the Pacific.

The S281 Course Team

Dave Adams
Jocelyn Bell Burnell
Cameron Balbirnie
Giles Clark
Alan Cooper
Sue Dobson
Carol Forward
Peter Francis
John Greenwood
Charlie Harding
Karen Hill
Jonathan Hunt
Tony Jolly
Barrie Jones
Bob Lambourne
Jean McCloughry
Elaine Moore
Cheryl Newport
Lesley Passey
Colin Pillinger
Ian Robson
Dave Rothery
Dick Sharp
Russell Stannard
Liz Swinbank
Margaret Swithenby
Ian Wright
John Zarnecki